At Home with Fragrance

At Home with Fragrance

CREATING MODERN SCENTS FOR YOUR SPACE

Kristen Pumphrey & Thomas Neuberger

Abrams, New York

Table of Contents

Introduction

We were walking along a sunny ridge of the Ewoldsen Trail in Big Sur when I smelled it. Sage. Not just any sage, but white sage that had been frying on the grassy slope of the trail just below. The salty Pacific Ocean glimmered in the midday sun. In front of us, the trail climbed upward before dipping back down into a cool and lush redwood grove. But all I could think about was the smell of that dry, sharp, sun-baked white sage.

It was a scent memory I took back to our studio in Los Angeles, locked into my brain the way you memorize your phone number as a kid. I kept thinking about this concept, of recreating the smell of that trail—sea salt and eucalyptus and redwood and white sage—and a year later, after revisions and sampling, we released a scent, Golden Coast, inspired by that moment.

But of course, you know, it wasn't just about the smell of the sage. We had long looked forward to our trip to Big Sur, not just because of its remote location, steep waterfalls, dense forests, and boutique hotels that we couldn't really afford. It was because we really needed some time off.

Just the year prior, my business had a huge break when our handmade candles got picked up by a major national retailer, catapulting us onto a roller-coaster ride of growth and expansion that consumed my life. Tom and I had also gotten married and he officially joined the business in the form of my Lead Candle Pourer. What started as an Etsy shop (but really a labor of love) in a second bedroom, had burgeoned into a business that, at the time, sold to hundreds of shops and several big retailers, such as Urban Outfitters and CB2. We employed people now—people that made the candles and shipped them and wrote sales orders right alongside us. After we got that big order, Tom and I worked for sixty days straight, taking only one day off, to get it all done.

We hustled because we didn't know how many more doors would be open to us—we saw this as our shot. I started P.F. Candle Co., (which stands for Pommes Frites—a play on my last name) in 2008, which was right at the beginning of the recession, and it was getting to a point where I struggled to make ends meet. I even considered giving up on my dream, which was pretty simple—to make a living by selling handmade stuff. But that eventual lucky break felt like a sign that I should take it and run, and although I would not recommend running yourself into the ground in quite the same way, I have zero regrets for the long hours and relentlessness that defined those early years.

So it was a pretty big thing for us when we chose to go away for a weekend. In retrospect, I think we only took one actual day off from work. We left our team of four in charge of the operations and set out on a trip to Big Sur, where we rented a tiny cabin and built fires at night, drank beer and ate organic food, read books by the riparian Big Sur River, and went hiking.

Why that moment was so important—the experience of hiking through all these ecosystems and smelling the flaky redwoods, the cool creek bed, the invigorating white sage—is that it was the moment we turned from being a company that sold handmade candles to a home fragrance company.

I learned to make candles at the age of twelve, so when I started my business, I naturally took them up alongside other goods. But candles were the thing that stuck, and over the years, I snowballed my scent intuition into actual training and conceptualizing, going from "something smells good" to "something smells good, and it's dry white sage." That trip was the first time I harnessed my own power of memory into a scent that would later become one of our bestsellers.

Scent is about so much more than just the way something smells. Scent is an experience. I think that's why I'm so drawn to this industry—I didn't want to just create a product that someone would use and be done with. I wanted to create a functional product that gave someone an experience that lasted longer than a moment.

Once I began to notice scents, my senses were heightened. The sense of smell is primordial; we learn to smell before we learn to speak. It communicates without words. It can tell you when something's good (like cookies) or bad (like a gas leak). Scent and the means by which you use it—lighting a candle or some incense—connect you to your space around you and create rituals that help soothe the stress of our relentless cultural pace.

It's not just that scent and fragrance can enhance your life and space, and make you more attuned to the world around you. It's also that having a deeper connection to how things are made gives meaning to the items you choose to keep around you. We wrote this book not just as a primer for you to understand why things smell like they do but as an homage to our roots—DIY.

Our little business has grown a lot since that hike in 2014. We are now a team of seventy, and we are sold in thousands of boutiques and chains worldwide, including our own shops in Los Angeles and San Francisco. This concept of DIY has driven everything that we have done. We grew the business with zero funding or business expertise, and were basically schooled on the internet and learned as we went.

At the tail end of writing this book, the entire world was hit with the Coronavirus pandemic. Overnight, our world changed, and we suddenly found ourselves inside for long hours at a time. I don't think I've ever burned so many candles in my life. I was panicked, thinking that everything we had worked for—the business we had so carefully built, the employees we had hired, all the candles we had made—would vanish overnight. But a funny thing happened: As we closed our doors, our customers found us online. They sent us DMs on Instagram letting us know how much a candle or room spray—a tiny piece of normal—meant to them, how it was helping them get through. They sent gifts to nurses and frontline workers, or just to their cousin who was working from home, to connect and break up the drudgery. Not only were they using these products as aromatherapy, they were shaping their interior life through scent and design. It was everything we had written about in the book coming to life. The products you choose to use in your home matter. They can comfort you, give you meaning, or give you a semblance of normalcy and control.

In the chapters ahead, you'll learn about what fragrance is and what it smells like. Then you'll see how to incorporate fragrance into your home design, or how to use it as an aromatherapy tool. Fragrance is in everything, from our dish soap to our bug repellent. Once you become aware, you can harness this to shape the world around you, and bring intentionality where there wasn't any before. And the fun part? Lots of DIY projects to get your hands moving. The connection to the things that you use on a daily basis—whether by learning more about the companies you shop from, or by making the products yourself—is something that will stick with you.

—KRISTEN

CHRIST · THE ALBUM
CRASS
SONY
TAYLOR SWIFT RED

What You'll Find in This Book

Part I

Fragrance Primer

This is an in-depth look at what fragrance is, and what fragrance actually smells like. Short of a scratch-n-sniff book, we've done our best to explain what our favorite notes smell like, and how you might use them. You'll learn about fragrance families and top/middle/base notes.

Part II

Fragrance & Design

This section explores the connection between fragrance and your home design. Here, you'll discover what scents work for which room, or which candle to burn depending on the mood you'd like to set.

Part III

DIY

In this section, we'll explore how to make all sorts of candles, hand-rolled incense, room sprays, potpourri, and more.

Fragrance Primer

If we can leave you with anything after reading this book, it's about approaching your home fragrance with intentionality. Kristen always talks about how in high school, her art history teacher set a group of objects out and went down the line, asking, "Is this art?" Some were obvious—paintings or sculptures of course classified as art. But what about the piece of driftwood? Not so much. What stuck with her was the importance of intentionality: The difference was that an artist created that painting or sculpture, but a piece of driftwood was just pulled from Mother Nature, and although it was beautiful, beauty does not equal art. (An argument can be made that by this teacher elevating this piece of driftwood in front of the class, the intentionality changed, but that's really more fitting for an esoteric dinner party discussion, not a book about home fragrance.)

The same can be said for fragrance. The difference between scent (the odor something gives off) and fragrance (an orchestration of scent molecules) is intentionality. Fragrance is something that is crafted, made by harnessing scent notes together. Fragrances can be built upon and improve the smells in your life. You get to choose your fragrance, which isn't always the case with general scents and odors. (Although you can take out the trash—that's a scent you can always choose to eliminate).

In this section, we're going to dive into the fine details of fragrance. We will take you beyond "Oh, that smells good" to give you a deeper understanding of the composition and texture of fragrances. This is a great base layer to have when exploring how to use these scents in your home (Part II), but you can also reference back to it and approach Fragrance & Design or DIY first.

WHAT IS FRAGRANCE?

Fragrances are just molecules that come within receptors in our nose and transmit data to our brain, creating a scent. All fragrance molecules are some combination of oxygen, hydrogen, carbon, sulphur, and nitrogen. Depending on the shape, composition, and orientation of that molecule, it will transmit different information into our brains.

The things you smell in your day-to-day tend to be complex combinations of compounds. Let's take a very popular scent and break that down: a simple lavender. Lavender isn't just one smelly molecule, however. It's made up of one hundred or so compounds, such as eucalyptol, camphor, linalool, coumarin, and lavandulol.

Oils that you use for fragrance are either going to be fragrance oils or essential oils. Fragrance oils are made by dissolving raw materials like powders, resins, and crystallines into carrier oils or alcohol. Many store-bought fragrance oils are not just one note but a complex mix of aroma chemicals, essential oils, and resins, known as an accord. You can choose to work with either synthetic or natural materials—we'll explain more here and which ones are our favorites.

SYNTHETICS

Synthetic fragrances are classified as aroma molecules that are not naturally derived but instead created in a lab. When people realized that fragrances were molecular compounds, they realized they could isolate the individual scents, and synthesize them. Synthetic chemicals may be isolated from petrochemicals, as they tend to have all the necessary elemental ingredients that can be rearranged into a fragrance compound, being derived from oil within the earth.

There are a few major benefits of working with synthetics: You have a vast network of notes to work from, the cost is more affordable, and they are very reliable and stable. The character of the fragrance will not change depending on where the material was sourced, whereas with naturals, the weather or where it was grown could yield a very different result.

Synthetics are highly regulated and because they tend to be more simple, from a molecular standpoint, they may be less likely to cause irritation (more on that in naturals). For us, we've always found it's a matter of dose, not the ingredients, that can cause irritation. When in doubt about an ingredient, scale it back.

Fun fact: The first isolated fragrance molecule was coumarin. It was isolated from the tonka bean, and it is a major part of tobacco and hay notes.

An issue with synthetics is that they tend to be very one note. The aromas we smell in nature are complex, so it can be very difficult to create an accord that really replicates the way you expect something to smell by using only synthetics.

NATURALS

Fragrances extracted from natural materials are considered "naturals." These could be essential oils, absolutes, or CO_2 extractions.

The first natural extraction method yields essential oils. Essential oils are extracted from a natural material through either steam distillation or cold press. This leads to fragrances that are lighter and fresher than those of other extraction methods.

A second extraction method produces absolutes. In the past, absolutes were created by putting natural materials in fats, which would extract the oils; then it would be washed with alcohol to remove the fats leaving only the fragrance oil. Today absolutes tend to be extracted using hexane. Hexane is a petrochemical that causes the oil content to leave the material. Absolutes tend to be stronger fragrances compared to essential oils, but are significantly more costly.

Another method of extraction is CO_2 extraction. Here, materials are placed in CO_2 and the pressure is changed, causing the oils to be extracted from the natural materials. CO_2 extraction is the newest method and the fragrances tend to be the most true to life.

The benefits of natural fragrance are that they are going to be closer to what you actually experience in life. The possible hundreds of compounds in a natural product can make it difficult to replicate with solely synthetic ingredients. That naturals are made of so many compounds is something to consider, as there may be unintended reactions because of chemicals you may not be aware of. For that reason, naturals tend to be highly regulated.

ISOLATES

Somewhere between natural and synthetic are isolates, which are single scent molecules extracted from natural materials. Simply put, these ingredients are extracted from natural materials, but are not essential oils or absolutes. They can come from a living organism, or even bacteria or fungi, whereas synthetics can be synthesized from inorganic materials, and naturals are derived via cold press, steam distillation, CO_2, and solvent extraction. Most suppliers will classify these as all-natural fragrances or fragrances from naturally derived materials.

A NOTE ON NATURALS VS. SYNTHETICS

In our practice, we use both synthetic and naturally derived fragrances. Many people assume that using essential oils is the "best" way to go, but we like to use them sparingly for a number of reasons. What it boils down to is cost, performance, sustainability, and artistic direction. Number one, the cost. If you were to fragrance your products with only essential oils, you'd be spending a *lot* of money. For one of our products, we went the "all-natural" route, meaning that the fragrances were from natural sources, but not made using essential oil extraction methods. Even then, our fragrance costs were double that of our other lines. Performance-wise, essential oils have a low flash point (meaning, they evaporate when exposed to heat), so you tend to lose a lot of those great top notes in the production itself when it comes to candles.

From a sustainability standpoint, it takes a lot of raw material to make just a few ounces of essential oil. For example, it takes 105 pounds of rose petals to make just five milliliters of essential oil. Another reason to use them sparingly! In cases like sandalwood, where sandalwood is an endangered species, it's actually more sustainable to use a synthesized version.

The last reason is purely artistic: When you work with man-made and natural, you have a huge library to choose from. I compare this to a painter limiting themselves to materials that can only be derived from natural sources, like turmeric or avocado peels: awesome for them, but not for everyone. Some people love the additional challenge of working with a limited palette, but it's fun to experiment with a broad palette first.

In terms of safety concerns, look for phthalate-free fragrances. You can also have extra assurance by looking for fragrance oils that meet guidelines set by the IFRA (International Fragrance Association) and the RIFM (Research Institute for Fragrance Materials).

The best fragrances are going to be the ones that are a blend of both natural and synthetically derived ingredients. These give the character of the fragrance body and heft.

FRAGRANCE FAMILIES

Fragrance types, or fragrance families, are a way to collect many different types of fragrance into comprehensible categories. In perfumery, there are a few widely used methods, including Michael Edwards's Fragrance Wheel. This is a comprehensive and in-depth approach to categorizing fragrance, but for our purposes at P.F., it can be a bit overwhelming. We created broad scent categories for our customers based on what they told us they were looking for, i.e., "I'd like something a little woody," or "Do you have anything fresh like the ocean?"

We almost never have a customer ask us for a "chypre" style fragrance, but that doesn't mean they don't like that type of fragrance—a little mossy and green, like a damp forest floor. So our approach with our customers, and here in this book, is twofold: One, explain it in a way that you'll understand if you never looked further than a perfume notes section, and two, educate so we can deepen the experience and appreciation for all things smelly.

We've broken fragrances into five categories here: earthy, fresh, floral, spice, and gourmand.

EARTHY

Our earthy category is a catchall for woodsy, straight-out-of-the-ground plants and trees. This category spans the gamut, from dry woods like cedar to earthy sandalwood, or warm resins. The woods family is P.F.'s home base: Many of our fragrances are built off a woody base, like sandalwood or patchouli, and layered in for complexity. Earthy scents are grounding, and tend to be heavy since they're composed of larger, long-lasting fragrance molecules.

WOODS

Woods have become increasingly popular among a certain type of fragrance company (guilty). It's easy to connect with the feeling of going for a hike among the conifers. We also include earthy patchouli and rich oud in this category, since they create the same texture within a fragrance as something like a cedar.

Cedar

Woody, clean, and astringent, cedar is very ambery. Cedar goes between smelling like fresh-cut wood to a hamster's cage. Cedar gives brightness to fragrance blends.

Sandalwood

Sandalwood is a building block for us. It's like a really fantastic jean jacket that goes with everything. It's woody and earthy, smooth like vanilla, green and a tiny bit damp like oakmoss, but the top is dry and bright like sawdust.

Vetiver

Vetiver is green and grassy and dirty; it smells like sharp damp hay and a touch of cigarettes. There's something very animalic about vetiver, so if you're looking to create a sexy scent that's still grounded and earthy, this is your note. Vetiver smells like roots, and is a little smoky like birch tar—it could work really well to create a whiskey vibe.

Patchouli

Patchouli gets a bad rap, but it's one of our favorite fragrance notes. Succinctly, patchouli smells like aromatic and invigorating dirt—in the best way possible. We often try to create scents that smell right out of nature, and patchouli is your friend for offering a dirty, earthy scent. It's got an amber quality, and smells clean and a little minty without being astringent. Great for scents that are meant to have a soil component.

Oud

Oud is a note you want to add to your fragrance if you want it to smell expensive—and it is. Oud is the resin from the tropical agar tree; its wood chips are sometimes burned as incense in ceremonies and celebrations throughout Southeast Asia and the Middle East. Oud is warm and sweet—rich, resinous, woody, and balsamic. An incredible material that gives fragrances depth.

Pine

Pine notes are green, woody, and dry. Unlike their Christmas-tree partners, fir notes, pine is not as sappy and sweet. Pine smells a bit cleaner and terpenic, which adds a lifted citrus aspect. Pine notes would pair well with fig or something balsamic, or you could add marine or sea salt notes to take this in an "ocean forest" direction.

Petitgrain

Petitgrain is the leaves and twigs of the bitter orange tree. It smells like dirty ground (in the best way possible). This is like a dirty, woody citrus scent. It has an ozonic quality that smells like the ground after a rainstorm. A really high-end scent, but don't pair it with gourmands or it will go all "cooked food."

Oakmoss

Oakmoss smells green, damp, and spongy—you can really imagine the soft furry moss when you smell it. Oakmoss is a fresh and clean note with a

touch of saltiness. The heart of this has a creamy vanillic aspect that makes this a very popular fragrance. Oakmoss pairs great with lavender, citrus, or heavy woods. A versatile tool in your fragrance cabinet.

Juniper

Piney, resinous, and a little fruity (like pear-apple). This obviously has a berry aspect that smells like gin (since that's how they make it). Juniper is great in a kitchen because it feels clean, natural, and woody, but still bright.

RESINOUS

Resinous scents, like sappy amber or incense, smell sticky and warm like the sap of a tree. They achieve the same concept as woods, in my opinion, while being a bit more robust and not as dry. Some of these notes can also be pretty powdery.

Amber

Amber has a powdery, tickling quality—you really feel this one in your nose when you smell it. It's sweet and effervescent, with a sticky and resinous heart. Amber lends brightness to scents and makes them drier.

Resin

A resin note smells like you're standing right next to the bark of a tree and smelling its sap. Alternately, there's a piney, sawdust aspect that's very balsamic—a little sweet and woody. It's warming and comforting. Resin should be present in your woody blends, as it adds a warm, rich quality that can make a scent less sterile.

MUSK

So, fun fact. A lot of people can't smell musk. Or, they think they can't. It's something like fifty percent of the population that is anosmic to musk—and up until a few months ago, Tom thought he was one of them. One trick to see if you're truly anosmic, taught to us at the Institute for Art and Olfaction in L.A.'s Chinatown, is to smell three strips—one dipped in musk, and the other two not dipped in anything. If you can detect a difference in one of those strips, you can smell musk, but you may need to continue to develop your nose for it. As you become aware of what it smells like, you'll be able to pick it up in other fragrances. For the other fifty percent, musk is described as heavenly and one of the best things they've ever smelled. There is something sexy, or rather sensual (sorry) about musk. It has an ozonic feeling, a little cold and wet, but one thing musk has is a lot of body. Musk smells clean, sometimes a little soapy, and sweet. Musk is intriguing—just think about what you would describe as musky on a person. Some people may use the term musky negatively, but for someone you're attracted to, musk is what draws you in. It's incredibly personal. Musk is a great additive in a laundry scent to give it some body.

LEATHER

Leather notes remind Kristen of going to a saddlery, since her sister was into horses. The rich smell of saddles, bridles, and boots can be polarizing, but it's a scent that's easily incorporated into home design. Leather is cool and can almost go a little minty-herbal. There's a sharp coumarin-hay aspect that needs to be balanced out. Great in small doses, but it could also stand on its own, mixed with something brighter like a citrus.

FRESH

Fresh scents are invigorating, bright, and citrusy. We consider citrus, aromatic scents (in the fragrance sense of the term), and marine scents to be fresh. Fresh scents have a light, airy quality about them, and are great to enjoy while working, or to make a room less stuffy.

CITRUS

Citrus is a scent that typifies California because of its agricultural history. Citruses are invigorating and clean. Sometimes citruses have negative connotations because of their frequent appearance in household cleaners. Stick with high-end citruses like bergamot or neroli, or use essential oils, for the most sophisticated application.

Bergamot

Bergamot is the scent of the bitter orange, typically found in Mediterranean climates. Because the essential oil is extracted from the rind, I always compare this to scratching the inside of an orange peel—that juicy and textural feel. Bergamot smells like citrus, floral, bitter soil, tea (it's used to flavor Earl Grey tea), and it's a touch terpenic. There's some black pepper spice in there; a short way to describe this scent is that it's a vegetal citrus. It's not sweet, so it doesn't create a candied effect for your blends.

Grapefruit

Grapefruit notes are juicy, and a little sour and bitter. It smells like the white fleshy parts inside the peel (the pith), and has notes of acetone and red fruit—like strawberry and cherry. It's full-bodied, so it can stand as the core of your fragrance, or just lend a little unexpected citrus lift. It's a great scent for a kitchen because it pairs well with cooking smells. Also a wonderful "pick me up" scent in the morning.

Lemon

Lemons are citrusy, sweet, sugary, and zesty. The smell of a lemon peel is great, as it's a bit more sharp and zingy. Use lemon sparingly so it doesn't go the "household cleaner" route.

Tangerine

Tangerine is a fun and unexpected citrus. It performs like an orange, but it's more sour and it has a light feel to it. Tangerine notes are juicy, and they would make a scent feel airy and energizing.

Orange

Orange is a full and robust citrus note, a touch bitter and sweet, but not as sour as some other citruses. Orange is great at the heart of your fragrance to build off of, rather than using it as an accent like other citruses. Orange has a great green, cut-grass (triplal) component.

Lemon Verbena

Lemon verbena is a bushy shrub usually associated with herbal healing or spas. It smells lemony, grassy, and a little floral all at the same time, in the vein of lavender with the linalool aspect. Linalool smells like it's on its way to being floral—but stops just short of the blooms. This is one of those scents that is frankly perfect on its own and needs no additional notes, but if you're looking to spice it up, you can try some basil herbs to deepen the green triplal effect.

Lime

Lime is a little sugary and of course always reminds us at best of margaritas—or at worst of bad cleaning agents. Lime is common in grocery store–bought cleaners. It's incredibly bright, so you just need a touch at the top for an impact.

GREEN

Green scents are freshly cut grass, conifer trees, foliage, vegetal scents. Green fragrances are an excellent partner for city dwellers who don't have space or enough light for a jungle of plants in their homes. I often think of green scents as juicy (like aloe vera) because they're fresh and alive, but also lush and damp. Green scents can have a bit of a preppy feel (maybe thanks to the signature scents of stores like Abercrombie).

Basil

Basil is a scent of happiness and peace. There are two varieties you can find in fragrance: sweet basil or Thai basil. The sweet basil is herbaceous, sweet, and great paired with a citrus scent. Thai basil is going to smell spicier and earthy while still retaining its green quality.

Cut Grass

Cut grass, or the smell of triplal, is a scent of childhood. It always smells a bit damp, and full of leafy, green chlorophyll. It's a fantastic springtime scent, or it could stand in if you're looking to create a "fern" scent, although, in actuality, ferns don't have much of a smell to them.

Sage

Full bodied, a touch sweet and aromatic; herbaceous in the lavender sense; great in an essential oil. A key tool for relaxation and calmness, and, like jasmine, can help alleviate anxiety.

Rosemary

One of our favorite natural smells (have you ever run your hand through a rosemary bush?), but it's difficult to replicate, even in essential oils derived from the plant. The best rosemary smells camphoraceous (woody and minty at the same time), herbaceous, and cooling—it opens up your sinuses. Rosemary is great in the office when you need a little lift.

MARINE

Marine scents reek of that salt life in the best way possible. Common marine notes are sea salt, ocean air, algae, and sea water. I didn't have much of a connection with marine scents until we started introducing them in our line. At worst, they can be astringent; at best, they're damp and lush and make you nostalgic for summertime.

Sea Salt

Although technically salt could be a gourmand note, we're putting sea salt in marine because of this note's ability to render one oceanside. Sea salt is reminiscent of ocean water, dried on your skin. It's piquant and bright and gives fragrances levity. Sea salt is mandatory in an ocean fragrance, but it also pairs well with green or herbaceous scents like rosemary and vetiver.

Rain

There are few things better than the natural fragrance of wet and ozonic rain. Rain notes capture not just the damp water but the feel of the atmosphere. If you're going one step further, you can create the scent of petrichor, which is what the atmosphere smells like after that first rainstorm of the season, by adding in earthy wood and soil notes.

AROMATIC

Aromatic scents cause a cooling, opening sensation when you smell them, kind of like Vicks VapoRub. This not only pertains to mints and eucalyptus notes, but is also a component of invigorating herbaceous notes like lavender and rosemary. Who knew.

Spearmint

Heavily associated with chewing gum (Wrigley's), this mint is very sweet and tangy.

Peppermint

Peppermint is hard and sharp. It can be sweet like a peppermint patty, but it's definitely full-bodied. You'll feel this one through your eyes and nose as the aromatic eucalyptus and clove notes clear you out. A great energizing scent that you can use when you're trying to capture the feel of the outdoors.

Eucalyptus

We're partial to eucalyptus because it's such a common foliage here in California. Eucalyptus is cooling, green, and a little oily—it's super invigorating when you take a whiff through your nose. You can use just a touch of eucalyptus, and it's great in marine scents to create that "cold Pacific Ocean" feeling.

FLORALS

Florals have traditionally been seen as romantic, and frankly, they're one of the hardest fragrance groups to nail. There are few things more heavenly to me than the smell of a fresh, blooming rose. We frequently make visits to the Rose Garden at Exposition Park and explore the varietals and see how they smell. White florals like jasmine and gardenia are another insanely popular group. Part of our job—creating unisex fragrances— is to debunk the gendered connection with florals and traditional femininity.

Lavender

Lavender is one of the most popular scents, and for good reason. Who wouldn't want to live on a lavender farm? (Or is it just me?) Technically, lavender smells like linalool, eucalyptus, and terpenes had a baby. In layman's terms, it's herbaceous, cool, and dank, a touch piney, citrusy. It heads toward being floral without the blooms (which is how I describe what linalool smells like). I categorize lavender under floral, but it could also go under aromatic or earthy.

Jasmine

Heavily indolic, which on its own, is not a nice scent (indol smells like decay and halitosis), but when composed with other components creates a robust floral. You see, things can't smell truly alive unless they are dying a little bit. Jasmine is heavy, a little sweet, with a touch of fruity banana. This is a scent that's hard to get right—a lot of people don't actually like the way straight jasmine smells. Balance it out with another white floral to create the "idea" of jasmine (the mind is a powerful thing).

Neroli

Neroli is a floral citrus, being the blossoms of the bitter orange tree. Neroli smells like a bright and spicy white floral. Great in upscale applications, paired with bergamot or something herbaceous.

Rose

A really good rose is soft like petals, with honeyed citrus and a sweetness that isn't cloying. It smells damp and green. You'll have to smell a lot of rose oils to find the one you like; the easier way is to smell rose accords that are pre-blended by perfumers. Rose has a little kick to it, with a fruity component like peach-pear. The "bad association" with rose—that it's powdery or stuffy—comes from bad accords, and is not the true character of the molecule.

Geranium

Geranium is floral, but has a bit of a citronella aspect to it. It's very terpenic and a little minty, with the traditional petal notes you'd think of. We'd use geranium to build a bright floral scent; it's great in cleaning products or in rooms you want to smell "clean"; I also love it on the patio.

Violet

Violet smells oily and green, almost like a cucumber. This is a floral that's not heavy on the bloom. There's a touch of orris root (the root of iris) on the dry down, which makes it heavy and powdery and a little bit like wet concrete. Violet is a great addition to a rainstorm scent.

Tuberose

White floral, fruity, coconut-y, with a gritty sunscreen feeling like monoi, and a touch vanillic. This is a favorite floral note, but it can definitely stand to be toned down in an accord.

Chamomile

Chamomile is a great minty alternative for an herbaceous bouquet scent—it gives depth and body. Roman chamomile has a damp, funky smell like the stems of flowers that you've left for too long in the vase. Great for a bathroom or kitchen. Chamomile can be calming, so it's also a good match for the bedroom.

Ylang Ylang

Very intense floral, with a peppery and anisic root-beer thing happening—that's balsamic for you. It's got a creamy top with papaya notes. Very heavy and heady—a good note to use if you want your floral to be "blooming."

SPICE

Spice scents add warmth and character. Spice notes are intriguing and tickle your nose, and will make you want to keep coming back for more. Add just a touch of spice to perk up a scent, or add a whole bunch to create a full-on holiday explosion.

Black Pepper

Black pepper is terpenic without being astringent—it's fresh and invigorating with a woody base. It is one of our favorite spice notes to use because it's spicy without being "holiday." It's great in small amounts in unisex fragrances, or paired with a floral.

Clove

A holiday classic through and through. Or, maybe it takes you back to when you were a teenager and secretly smoked clove cigarettes (just me?). Great with orange, like in a pomander, and would also pair nicely with a rose.

Cinnamon

Hot, balsamic, a little sharp. Cinnamon is most definitely a holiday scent. One of our favorite markers of the season is when they bust out the cinnamon brooms at Trader Joe's. Great in small doses with apple, sandalwood, or woody notes.

Cardamom

Cardamom is peppery, citrus, and evergreen; it has such a breadth of notes that it can stand in for more traditional clove or cinnamon notes. Cardamom has a full and robust earthy quality.

Ginger

Ginger has a green and bright, almost citrus characteristic to it. It's a great spicy scent to use because it's not straight from your traditional dried-spice cabinet. Ginger is warm and fresh, and is a key tool when you're trying to bring focus to your space.

Myrrh

Myrrh is one of our favorite "spices." It's pungently sweet and resinous—truly balsamic with woody and sweet notes at the same time. It pairs wonderfully with something herbal, and is a preferred note to use in incense.

Coriander

Coriander is the seed of cilantro, so this scent is spicy but retains a little bit of that linalool-green quality of the plant. Coriander is earthy and nutty, and is a welcome break from the cinnamon-nutmeg spices.

Caraway

This seed is traditionally used on rye bread, so it's a delicious association. The seedy note has a green aspect similar to dill, and is great in the kitchen, or in a woody blend.

GOURMAND

Gourmand scents bring out the snob in many people—the association with fall scents and the ubiquitous pumpkin spice makes eyes roll. To that we say, just give in. The gourmand category is straight fun; it's the pop-culture gossip of your fragrance tool kit. Strawberry scents in particular are summery, and youthful; paired with a chamomile and sandalwood, it would be a grown-up Midsummer vibe.

Red Fruit (Strawberry, Raspberry)

A lot of red fruits tend to smell the same, so to save you from reading a similar description over and over, we've lumped them into one category (don't worry, a professional perfumer signs off on this approach). Red fruit scent is sweet berries that are juicy, and is evocative of the way dark pink would smell. We use a touch of a raspberry note in our Campfire scent to give that sweet, natural s'mores quality.

Green Fruit (Kiwi, Pear, Lychee)

Green fruit notes are sweet, with a jammy texture that creates lushness in fragrance blends. Green fruit gives a dense and fruity body.

Currant

Currant smells like sweet, ripe fruit. It's a very youthful scent; it actually reminds me of a Bath & Body Works lotion popular with teenagers. It's a good note to use for fruit details that doesn't go too gourmand.

Apple

Apples are juicy and very sweet, with a touch of bitterness from the rind. Apples also have a fleshy, green quality to them.

Vanilla

Vanilla smells like cake batter, but a little woody. It is powdery and sugary. Vanilla is in so many of our foods, that it's hard to say if vanilla smells like something, say cocoa, or if cocoa smells like vanilla. Vanilla is a base note that will last a long time.

Coumarin

Coumarin is a major constituent of tonka bean, hay, and tobacco. It has a strong balsamic scent—sweet and spicy. Smelling coumarin is like smelling a woody vanilla cake. Sweet and savory at the same time.

Pumpkin

A real pumpkin note, and not a pumpkin spice note, smells like the pith of a lemon. Ripe pumpkins always smell like they're almost *too* ripe. Blending vanilla, nutmeg, and clove notes with pumpkin will create a pumpkin pie situation.

Fig

Figs smell ripe and juicy and a little seedy. Imagine that moment when you tear apart the soft flesh of a fig and reveal the seedy belly—figs have a strong place association for us and remind us of our backyard in East Los Angeles, and eating black figs off the tree—before the Japanese beetles got to it.

Pineapple

Pineapple smells sweet and acidic and juicy. It would smell great paired with a heavy, resinous base. Use it sparingly, though—it can be divisive.

TENACITY

aka Top • Middle • Base

Most people will recognize top/middle/base notes from perfumery, but what do they actually mean? These terms refer to the tenacity, or longevity, of a fragrance, which relates to the volatility of the materials. The more volatile, the quicker you smell it and the quicker it disappears. Simply put, top notes are light and burn off quickly (the way your perfume wears off), but base notes are those heavy notes that linger. We like to think about this in relation to music where you have your high, middle, and bass. A well-rounded fragrance would have notes from each—top, middle, and base—just as you might want range in your music. Tenacity is personal and subjective. What can be smelled for a long time by one person is gone in an instant for others. Some notes also blur the lines, ranging more mid-top (lavender) or mid-base (spices like ginger).

VACAY
ODE
Karter
500ml
APPROX.
400
300
200
100
500ml

TOP NOTES

Top notes are going to be light and volatile, so they tend to evaporate quickly. These are important because they give the first scent impression. In candles, we compare the top note to a "cold smell"—what you smell when you just smell the cold wax. From our experience, top notes tend to be citruses, fruit, and some florals.

PINE	SEA SALT
PETITGRAIN	CARAWAY SEED
PEPPERMINT	GRAPEFRUIT
SPEARMINT	LEMON VERBENA
LEMON	BASIL
LIME	CUT GRASS
SAGE	RED FRUIT
ORANGE	PINEAPPLE
BERGAMOT	EUCALYPTUS

MIDDLE NOTES

Middle, or heart, notes are released after the top notes wear off. These notes are robust and create the "body" of the scent. In candles or incense, middle notes come out when you burn the product. Common mid-note materials are florals and spices.

JASMINE	VIOLET LEAF
ROSE	TUBEROSE
LAVENDER	GERANIUM
CURRANT	ROSEMARY
GREEN FRUIT	JUNIPER
APPLE	CLOVE
FIG	CARDAMOM
PUMPKIN	CORIANDER
NEROLI	CHAMOMILE
TANGERINE	CINNAMON
BLACK PEPPER	

BASE NOTES

Base notes tend to be woody or resinous. Base notes are the freakin' best! They linger long after the burning is done, and base notes are key for the robust, full texture of a fragrance, of someone saying, "Oh, it smells so good in here!" Base notes are mysterious and warm, and are associated with "dry down" in perfume. Fragrances heavy in base notes tend to smell rich and luxurious. You'll notice a lot of wood and resin materials in the base-note category.

SANDALWOOD

CEDAR

LEATHER

VANILLA

RAIN

ORRIS

AMBER

OAKMOSS

OUD

MUSK

VETIVER

PATCHOULI

RESINS

COUMARIN

GINGER

MYRRH

YLANG YLANG

ACCORDS

Now that we've gone over notes, let's talk about accords! Accords are a set of notes that are pre-blended together to create a new, harmonious composition. I've often seen this compared to the way music is arranged: single guitar notes and piano may sound good on their own, but together, they make a melody.

Most fragrance oils on the market are going to be accords. When you are searching for fragrance oils, pay special attention to the note descriptors. You'll have to do a lot of smelling to figure out what you like and don't like—descriptions aren't always accurate. When I first started blending my own fragrances using commercially available oils, I could only find accords, so at first I found it tricky to create that custom scent exactly. I got in the habit of ordering one-dollar fragrance samples just to smell things—truly, until they invent scratch-n-sniff, that's the way you've got to do it! If you smell a lot of things, you'll start to pick up on the individual notes within an accord.

The benefit of accord fragrance oils is that if you just want to make some quick candles or diffusers, they've done the hard work for you. If you want to go one step further and create something custom, just blend two accord fragrance oils together. These create truly complex and beautiful fragrances.

Many fragrance websites will list their fragrances by type. For example, you can search for marine- and woods-style scents and find a bunch of accords in that vein. I think a more dynamic way to search is by narrowing down the specific note that you want. Often, our fragrances start with a simple note. In 2015, we became obsessed with the idea of creating a eucalyptus candle, because it reminded us of our first California home in the beachy Bluff Heights area of Long Beach. There was this massive old eucalyptus tree that hung over the 1960s apartment complex pool. From there, our inspiration went to other classic California scents, like orange. We searched supplier websites for all the eucalyptus and orange fragrances we could find. One note that popped up for orange was neroli—a honeyed, orange blossom take on the citrus fruit. We bought both accord fragrance oils, blended them together, and after many revisions, No. 16: Neroli & Eucalyptus was born.

Part II

Fragrance & Design

PICKING A FRAGRANCE FOR YOUR HOME

Fragrance is a crucial, but sometimes overlooked, component of spatial design in your home. Hotels pay thousands of dollars in marketing development to create signature scents that are pumped through HVAC systems, crystallizing guests' memories of the space. Even Disneyland uses scent to enhance their guest experience. As a good Southern Californian, we make biannual pilgrimages to the famous theme park. My favorite time to go is the holidays; it's just an incredible magical explosion of lights and holiday cheer. Ever since I've been making these holiday trips, my favorite ride has been "It's a Small World," that iconic Disney journey through cultures around the world, in no small part because of their use of fragrance on the ride. I remember the first time I ever smelled it—Tom and I took a day off work in December of 2014 to celebrate his birthday. This was a pretty much unprecedented event, because ever since the business started to grow in 2013, we had been working at breakneck speed and rarely took time off (not recommended, FWIW). As the ride made its way through technicolor Christmas lights, I smelled something. And I recognized it. It was the smell of a gingerbread house—baked goods, candy, and a touch of nutmeg. Because of my work, I was intimately familiar with this particular scent—we used a similar, if not the same, fragrance for one of our candles we called "Sugar and Spice." I grabbed Tom's arm gleefully—"It's Sugar and Spice!"—and to this day, "It's a Small World" is one of my favorite experiences.

This past holiday season, I noticed they went a step further. Every room was scented to match the decor. Hanging, prismatic lanterns and sparkling poinsettias smelled like cinnamon, and the Hawaiian room with dancing Lilo & Stitch smelled like coconuts and mistletoe. While we walked toward Sleeping Beauty's castle afterward, it started to snow (in reality, bubbles came out of the sky, which is the closest Anaheim is ever gonna get to snow) and I could have been imagining things, but I swear those bubbles were fragranced. Disneyland really stepped up its game.

So why is it important? Disney, the King of Marketers, used scent in their rides and fake snowflakes to further enhance the experience—actually, to turn on all the senses. Disney wants to fuse these memories into kids' brains: "Remember that time it snowed at Disneyland and it smelled like gingerbread?" Now the next time they smell gingerbread, they'll think back to that ride and beg their parents to go back.

Scent and its link with memory has been deeply explored and is endlessly fascinating—to me at least. When I travel, I always try to take back scented products as souvenirs. This all started in 2013 on our honeymoon. We visited Paris and Amsterdam, spent our days walking, eating bread, drinking wine, and sweating (it was July, after all). Around this time, French pharmacies had reached a trendy fever pitch on beauty blogs, and I was anxious to get my feet in there. I also noticed something: French women smelled good. They walked around in clouds of this particular perfume, something warm and floral, definitely feminine, intoxicating but not overpowering. One trip to the pharmacie showed me what it was: Nuxe, a multipurpose dry oil that smells of florals like monoi and orange blossom, vanilla, and rich oils. Now, probably not all French women are dousing themselves with this before they leave the house, but certainly all the ridiculously good-looking and stylish ones were (not unlike Le Labo's Santal 33). I scooped up a bottle all those years ago and still have the damn thing, with just a minuscule amount of oil left. This type of souvenir hunting has a twofold purpose: one, to find a useful and not wasteful product to take home, and two, to use the power of fragrance for nostalgia, which as an avid memento-keeper, I am very fond of.

We've all been in a situation where a sudden whiff of sunscreen—gritty coconut and orange blossom—will remind you of a summer trip to the beach. You practically hear the seagulls while you take it in. Or maybe you open an old purse or book that once belonged to a grandparent, unsealing a time capsule of fragrance. This moment of discovering a scent you haven't smelled in so long, something distinctive and evocative of time and place, is like a little story for your mind. It's relaxing, sometimes sentimental to the point of melancholy, but always welcome. Scent is shorthand for your mind.

Not only do scent and your memory weave together, but scent has a powerful effect on your mood. Whether it's just a ritual and act of lighting all the candles, or the fragrance of the diffuser itself, fragrance is mood-altering in the same way colors are mood-altering. They can calm you, uplift you, make you feel cozy, remind you of good times, take your focus off problems.

Perfumery is somewhere between chemistry and art: The notes are meant to evoke reaction and feeling, just like a painting or sculpture, but how you combine them and make them is nothing short of science. If fragrance is an expression of yourself, artistically speaking, it makes sense as the final extension of your home design. The scent of your home when you first cross the threshold into your house makes you feel comforted. There is no better feeling than that.

Fragrance is all around us, from taper candles lighting a dinner-party table, to laundry detergent and hand soap. There's even fragrance in insecticides like Raid. Once you start to understand and notice the nuances of fragrance, you can understand how to harness all these scents to create a harmonious experience for yourself.

We receive frequent requests from Airbnbs and hotel chains for custom projects—a scent that they could burn or use to fragrance their spaces and subsequently sell in the gift shop, so people can take home a little piece of memory with them. One of our favorite Airbnbs, the Joshua Tree House, with locations in Joshua Tree and outside Saguaro National Park, use our candles in their spaces, but they also use scents of the area to fragrance their homes. In the shower, they hang creosote branches as a desert-inspired shower steamer; when the steam starts activating the bush, it smells just like the desert after a rainstorm. That was such a fantastic experience as a guest—I still think about it—and anytime I see the bush in the wild I'll be reminded of that. I also love that they incorporated something so geographically specific; in the age of the internet flattening cultural and local specificities, it felt really special.

One trend we've noticed is how much people love to customize. As a brand, while we don't create custom scents, we always recommend that people customize their space further by mixing two scents together. For example, during the holidays, Spruce and Apple Picking are amazing together. This is where our love of DIY projects comes in. It's not feasible to create one hundred unique items, but if you make it yourself, no one will have one like it. This is what has drawn me to DIY time and time again—from making my daughter a little dress with a Peter Pan color and sixties-style pink daisy printed fabric, to painting a chair that I found on the street and recovering the cushion. DIY is for the individualist.

Ahead, we're going to explore how to fragrance your home: what types of products work in what rooms, what scent is best in the bathroom or kitchen, and layering these for maximum ambiance. The customization of your home's scent will also rely on your own memories; you can use travel or childhood memories as a springboard for creativity.

—KRISTEN

We'll be exploring three ways to approach fragrancing your rooms:

1

Fragrance by room purpose, or letting the function of the space dictate the scent.

2

Fragrance for aromatherapeutic benefits, or picking scents based on the energy they will embed into a room.

3

Fragrance by decor, or thoughts on incorporating fragrance as functional home decor and an extension of style.

THE BELL JAR
STEINBECK
OF MICE AND MEN
CALIFORNIA
SIN and SYNTAX
CARL SAGAN
COSMOS
MARC MARON
THE GIRL WITH THE DRAGON TATTOO
STIEG LARSSON
THE ROAD
JOHN IRVING
HENRY ROLLINS
LEARNING THE TAROT
TWYLA THARP
DECAMERON
You Can Train Your Cat
WOMEN

FRAGRANCE

by

ROOM PURPOSE

Fragrance, like the design of your home, is meant to showcase your individual style. I've had the experience of seeing someone's home on Instagram and thinking it looks like a mail-order, cookie-cutter living room. The same for fragrance—instead of just purchasing whatever candle is on sale, you can use fragrance to convey a message to your family and friends and help shape their experience of your home. Good design is aesthetically on point, but great design is custom and individualistic.

My goal when someone enters my home is for them to say, "Oh, it smells good in here." Frankly, I wouldn't be doing my job if that wasn't the case. Fragrance design is an artistic expression of your overall home design, not just a means to an end to make something "smell better."

I'm going to walk you through the rooms in our home and explore my intentions of fragrance choice (type and family) for each room so that you better understand the choices that you can make. But remember—this is only an example, a template to springboard your own fragrance design.

HOUSE

Living Room

WHAT OUR LIVING ROOM SMELLS LIKE

CEDAR / SANDALWOOD / PIÑON / TONKA BEAN
BLACK CURRANT / SEA SALT

TYPES OF FRAGRANCE TO USE IN YOUR LIVING ROOM

For all-around fragrance:
DIFFUSERS

For spot treatment on your linens (hello, pets):
ROOM SPRAYS

On Sundays, and before guests come over:
INCENSE

For movie night, or every night:
CANDLES

The living room is most people's first impression of your home. As an adult, I've never lived in a house that didn't just open right into the living room. Just like textures, furniture choices, and artwork communicate your design choices, so does that first big inhale. Is there anything better than returning home from a long trip, taking a whiff, and saying "smells like home"? Coming back after time away from your home allows you to smell it the way a guest does. Whenever we return from a trip, I'm struck by the notes of wood and cedar, dust and soil, maybe a touch damp if it's been raining. On the flip side, taking that first big inhale and realizing you forgot to take the trash out after a week is a bummer. Better break out the incense.

Something to consider with your living room is what scent your furniture, home, and decor is giving off. Real wood furniture, leather couches, cowhides, linen, bouquets, dried arrangements, plants—these are all contributing to the overall scent. Our current home, a 1908 Craftsman, always smells a bit woody because of the redwood built-ins, even though they've been painted over. Our Craftsman is also mustier than our last house, an East L.A. bungalow built in the 1960s—it's got that "old house" smell. We don't mind that scent, as it's like patina for fragrance, but it does factor into the overall aromatic landscape.

For me, the living room is a blank canvas. I want a grounded feel in my living room since we unwind with low-stakes trash TV at the end of the day (mostly reality, although we recently indulged in a decidedly non-trash marathon of *The Sopranos* from start to finish), but I still want it to feel like a fancy yoga studio. You gotta have that hi-lo vibe, right? For grounding fragrances, heavy base notes like sandalwood and patchouli do the trick. These traditional base notes linger in the air for a long time, so you'll always have a little fragrance going. We also love straight-from-the-forest tree notes like cedarwood, pine, and spruce. As avid hikers and people who frequently want to disconnect from technology, these forest-y notes are a shorthand for my mind to the wilderness that relaxes me the most—like the Ansel Adams Wilderness in the Eastern Sierras.

In addition to woodsy scents, our living room always has a rotation of whatever new product we're developing. This keeps it fun and interesting. Woodsy notes are the base, a constant or fixed control element. When we burn new fragrances on top of that, we get a deeper understanding of how those fragrances will pop and play with other elements of our line. Right now, we have a Piñon reed diffuser as a base level. I flip the reeds every couple weeks (or, to be honest, when we remember to do so). The cedar and vanillic aspect really sings when we've been away for a while. Lately, we've been combining Piñon with our new scent, Swell. Swell is all things beach day without being too marine (marine scents are aquatic and damp). It's a little fruity, salty, with a thick layer of sea moss that gives it a vibrant, green feel. We love this scent in the summertime, when burning it transports us to playing hooky and taking the whole family to the beach.

We also rotate scents seasonally to pop off the base. From October to December, we'll throw in a few classic gourmand and balsamic notes. Anything that gives that fall feel—crunchy leaves, apple cider, crisp air, baking spices—those are A-OK in this East Coaster's heart. Uncommon fruity notes, like fig and currants, mixed with a spicy clove, are a great alternative to the traditional pumpkin.

Layering scents in the living room gives it depth. It's the difference between a couch, and a couch with a few throw pillows and a handmade blanket on it. When you layer fragrances, you may also discover new elements and levels you hadn't before, like the way sandalwood interacts with citrus. For a home fragrancier, this is product and fragrance research in and of itself.

Dining Room

WHAT OUR DINING ROOM SMELLS LIKE

FIG / AMBER / PALO SANTO / CLOVE / BEESWAX

TYPES OF FRAGRANCE TO USE IN YOUR DINING ROOM

For all-around fragrance:
DIFFUSERS, POTPOURRI

For dinner time:
BEESWAX TAPERS OR TEA LIGHTS

Here's where we come to the trickier aspect of home fragrancing: when there's food involved! When you're eating, you really want to smell the food you cooked (or Postmated), not a candle, which heightens your senses. The reality with dining rooms today, though, is that they rarely serve only one simple function like eating. Just like our minds are multitasking so are our spaces, and dining rooms function as a place for family dinners as much as a desk, craft station, photoshoot backdrop, and, in reality, a landing zone for bills.

When you're burning candles during dinner, you're going for ambiance, not fragrance. You're going for mood. You're going for ritual. Tom got really into meditation, and after reading about it, I understood how a practice like that helps you rewire your brain. Rituals can do the same. The act of just stopping for a moment and lighting some taper candles says, "Hey, we're here now. Let's put aside our day for a dinner." Dinners for us are about six minutes long because we have a preschooler who "wants to play!" And when she gets old enough, she'll graduate from just putting forks on the table to joining the ritual of lighting some candles.

The duality of the dining room is, therefore, an interesting place to fragrance. Consistent fragrance is a really nice mood booster, especially if your dining room is like mine and it's a "flow-through" room between living spaces and the kitchen. Consider this your bridge then. You'll want to fragrance your dining room with a diffuser, for all-around scent, and accent it with candles for mood.

In my home, I keep a diffuser tucked into my record cabinet, usually something woody with a touch of spice like our Black Fig or Dusk candles. These types of fragrances—woody, resinous, a little sweet and spicy—are called balsamic. Let's be real, there's a high probability I like these type of scents in my dining room because the word balsamic makes me think of salads, but in reality, balsamic notes work because they're grounded enough not to be cloying—and the spiciness matches in a food setting. Balsamics can be a little sweet and spicy, but they're never sharp; they're soft, dewy, and balmy.

BALSAMIC NOTES	
	RESIN
	LABDANUM
	OUD
	BENZOIN
	FRANKINCENSE
	BIRCH TAR
	ELEMI
	AMBER
	VANILLA
	TONKA BEAN
	BALSAM FIR

These materials give a rich and soothing spicy quality to fragrances. They're not sharp, but warm and enveloping.

The second aspect to fragrancing your dining room is during a meal. I'm a big fan of taper candles. When we sit down to dinner, along with remembering to put the napkins out (four nights a week, if we're lucky), we light some taper candles. Beeswax is really key to me here. Beeswax has a wonderful, naturally honeyed scent. It smells sticky, warm, sweet, and earthy. Remind you of any balsamics? See, we're creating patterns here.

In a pinch, we do paraffin tapers (we picked up some fun ones at one of our stockists in London, Search and Rescue, in an array of vibrant colors; frankly, paraffin just holds color better than other waxes). But I don't love the plastic smell that paraffin gives off, and once you extinguish them, you have to leave the room quickly. That last poof of plastic is not great.

Burning beeswax tapers are great for mood, and the light fragrance doesn't distract from the aromas of your food. I avoid heavily perfumed fragrances in my dining rooms for this reason. If you're burning a really strong candle and eating at the same time, it makes for a confusing experience, where suddenly you're eating jasmine-sandalwood-spice. Nah, we're good.

Shaws
Original
Since
1897

Kitchen

WHAT OUR KITCHEN SMELLS LIKE

WHATEVER MEAL WE COOKED LAST

FRESH AIR / WOODY INCENSE

ROSEMARY AND LAVENDER CANDLES

LET'S BE REAL—SOMETIMES IT SMELLS LIKE THE TRASH WE FORGOT TO TAKE OUT

TYPES OF FRAGRANCE TO USE IN YOUR KITCHEN

For all-around fragrance:
CANDLES, SIMMER POTS

For Saturday mornings, or "Oh shit, I left the bacon on" moments:
INCENSE

For a little pick-me-up:
ROOM SPRAY

NOTES TO ENHANCE YOUR KITCHEN SPACE

- ROSEMARY
- LAVENDER
- BASIL
- CHAMOMILE
- SAGE
- THYME
- MINT
- LEMON
- GRAPEFRUIT
- TERPENIC FRAGRANCES, LIKE INCENSE AND WOODS

The kitchen is an even trickier fragrance battleground. Remember being a kid and going over to a friend's house for the first time and taking in that scent? It had a lot to do with the type of food they cooked and how they cooked it. The kitchen scent is such a great part of your home's fragrance DNA, so don't shy away from that. But sometimes, let's be honest, it can be a little stinky. The approach in the kitchen should be based on whether you're trying to enhance or envelop—do you want just a pick-me-up, or are you desperately trying to get bacon smell out of your home before a one-year-old's birthday party? On a Sunday evening after I cook roast chicken, it smells amazing. But come Monday morning, I'm ready for that scent to go quietly into that good night.

Enhance

I love natural fragrances in our kitchen. The organic, light quality really plays into enhancing food smells rather than being a cloying mess that's attempting a cover-up. If you don't go the full "essential oil candle" route, just pick scents that are plant-based for the lightest approach. Green scents, in the herbaceous and linalool-inspired vein, are great, as well as fragrances that lean clean, fresh citrus.

Envelop

There's been a time or two when I make something that just tends to linger. I'll never forget my niece's first birthday party: My sister had prepared a baked potato bar, but was stressing moments before guests were about to arrive because the house reeked of bacon. We threw open all the windows—which is a trick I use not just in the kitchen, but everywhere, to move air (and fragrance) around the house—and I vaguely remember it was someone's job to just wave bacon grease smoke out of the house. (I also want to note that a lot of people like the smell of bacon, myself included, and this was 2015 when woodworking, mustachioed, flannel-wearing guys were still really popular. I tried to reassure my sister of this, but I think she wanted to go a more subtle route for this occasion.)

So, if you're in a "bacon before the party" situation, in addition to opening your windows, here are a few tricks:

INCENSE

Incense is a Saturday morning staple at our house. Sometimes we'll do a little brunch and have friends over, but usually Saturdays are reserved for slowly imbibing copious amounts of warm coffee and grazing on pastries and fruit. Tom opens the windows and lights some incense, whatever we're into, so we can actually enjoy the fragrance on a slow weekend morning. Incense is awesome for quickly erasing bad smells. I'd look for something that isn't too sweet or marine—those types of scents don't typically blend well with food smells.

LEAN IN

Light up a fruity, gourmand scent and really lean into that bacon smell. What type of scent would pair well with what you're cooking? Look at your ingredients. I think herbaceous fragrances do especially well here because they round out that "spice cabinet" feel.

Bedroom

WHAT OUR BEDROOM SMELLS LIKE

CEDAR / SANDALWOOD / JASMINE
SMOKE / VANILLA

HOW TO FRAGRANCE YOUR BEDROOM

On your sheets:
ROOM AND LINEN SPRAY

On your dresser:
REED DIFFUSER OR ESSENTIAL OIL DIFFUSER

On your nightstand (blow it out before bed, k?):
CANDLE

Bedroom fragrance has the opportunity to be very personal—think about it like your secret perfume scent. This is your judgement-free zone, where you're creating a fragranced space just for yourself, or yourself and your partner.

In terms of base layer scents in a bedroom, they tend to be a bit musty, with old clothes and sheets with dried night-sweat in them (sorry, but it's true). We try to keep the windows open as much as possible, but in certain areas—whether it's due to safety or pollution or weather—that isn't possible.

The VIP of our bedroom fragrancing is linen spray, to keep sheets fresh. However, candles are a frequent nightstand companion in our customer's photos, so their mood-setting ability can't be overlooked.

So yes, candles are for ambiance in the bedroom. You're going to want them to be a bit sexy, but everyone's version of that is different. We'd steer clear of lemon or orange citruses, or anything too bright. Bergamot is a good citrus option that feels unisex and not like a candy or bathroom cleaner. Bonus points that it's a note in Earl Grey tea—that link means bergamot automatically relaxes me just like a cup of tea would do.

I love musks and heavy woods in the bedroom, like sandalwood, white musk, amber, and cedarwood. I pair these with a white floral like gardenia or jasmine, or maybe a golden rose, for a touch of romance. If you're more at peace hiking through the woods than relaxing in a spa, try oakmoss. A traditional chypre accord might even be up your alley, which has been described as smelling like a damp forest floor in autumn.

Jasmine has an added benefit, like lavender, of promoting relaxation. Jasmine has even shown to be an anti-anxiety scent, which if you're an insomniac like Tom, can be extremely helpful.

Room sprays are such a great way to incorporate fragrance into your bedtime routine. When our daughter was first born, we started using Fat and the Moon's Calm Kid Mist during her bedtime routine, and we still use it to this day—she won't go to sleep until she's had some sprayed on her hands, which she immediately takes a huge whiff of. It has notes of lemon balm, lavender, and poppy tincture.

Essential oil diffusers and room sprays are great in the bedroom because they're low maintenance—no need to worry about falling asleep with a lit candle on your bedside table or dresser. Plus, their mist is atmospheric and soothing.

FRAGRANCE FOR A CALM BEDROOM	LEMON VERBENA LAVENDER JASMINE
FRAGRANCE FOR A SULTRY BEDROOM	AMBER SANDALWOOD CEDARWOOD GARDENIA
FRAGRANCE FOR A PLANT-INSPIRED BEDROOM	CHAMOMILE OAKMOSS LINALOOL ROSEMARY

Bathroom

WHAT OUR BATHROOM SMELLS LIKE

DRY PINE / TEAKWOOD & TOBACCO
PEPPERMINT (DR. BRONNER'S) / LAVENDER
HERBACEOUS FLORAL / SOAPY

HOW TO USE FRAGRANCE YOUR BATHROOM

REED DIFFUSER
Your strongest, please. I do my diffusers in a dry woodsy scent like Piñon or Teakwood & Tobacco.

ROOM SPRAY
For taking care of business after you take care of business (sorry). Leave it on the back of the toilet for guests, and hopefully they get the hint.

CANDLES
For ambiance, for testing out new scents, when people come over, when it's Tuesday.

INCENSE
Odor damage control—in use of emergencies only, because the small space of bathrooms + lots of smoke is not a fun combo.

POTPOURRI
It looks nice and it lightly fragrances your space.

Ah, the bathroom. This is probably the number one spot for fragrance as a masking tool, but this is a tricky area to fragrance because once you use a specific scent in a bathroom, you may never un-smell it that way. Take, for example, another type of bathroom: the cat litter box. In a lot of houses, the bathroom also doubles as the cat box room. In our old house, we converted an office to a dual purpose office/litter box room (I bet you can guess how much work actually got done in there). We used a Patchouli Sweetgrass reed diffuser to mask the odor, and it did an incredible job. Too incredible, really—now the scent of litter box and sweetgrass, sweet and sharp, will forever be intertwined. For this reason, we refer to this process as "sacrificing" a scent.

Marine and lemon scents have long been a martyr to affecting a vision of "cleanliness" in bathrooms or other tiled areas. When you smell a particularly marine scent, like an Ocean Breeze candle, or something in the Mr. Clean variety, it doesn't necessarily inspire a spa retreat. It just reminds you of that Saturday you spent scrubbing the floor.

When fragrancing your bathroom, first think about the type of environment you want to create. For our house, there are two modes: function and relaxation. The reality is in most city apartments, or old houses, there's not a lot of room for the latter when the former is the most important. Scent shapes the shortcuts in your mind for "This is where I unwind and take care of myself."

Especially on days where I feel overwhelmed or a little down, a shower or bath is a way to respect myself. I'm ritualistic when it comes to baths in particular; there's a specific incense I like to light in the beginning (Australian Native incense from Addition Studio, which smells earthy but has levity and life, with eucalyptus and acacia), marking the start of the ritual. I open the windows and let a breeze come in, then I run the bath, scooping in a healthy cup of Epsom salts, argan oil, and maybe a "bath tea" of sorts, usually with rose petals and Himalayan salts. On the side, I light a few candles and have a cup of tea.

When it comes to functional fragrance in your bathroom, you're looking for scents that have some power. I wouldn't recommend vanilla-style gourmand scents in the bathroom, but citrus, oakmoss, fruit, and floral are all on the stronger side. Strong woods scents—not balsamic like a Christmas tree, but invigorating like campfire or teak—smell like a breath of fresh air in the bathroom.

Bathrooms are also a fantastic place to try out new fragrances—the small space and lower ceiling heights in the bathroom will give you a more accurate depiction of the strength, character, and performance of a product. There's fun to be had with natural elements in the shower itself. For cold and flu season, a bunch of eucalyptus tied around the shower head creates a steamy and invigorating adventure when you shower. Plus, it adds an organic-rustic feel to your bathroom.

Another option that toes the line between fragrance and medicine is shower steamers. With just a couple ingredients (baking soda, citric acid, witch hazel, essential oils) placed in the bottom of your shower, the steam from the shower creates a giant essential oil diffuser environment. Very atmospheric.

Epiphone
SMELL YA LATER
RECORDS
DSL
40C

Office

WHAT OUR OFFICE SMELLS LIKE

WHATEVER THEY'RE POURING IN THE FACTORY

HERBAL NOTES, LIKE ROSEMARY AND LAVENDER SANDALWOOD

GREEN NOTES—A BYPRODUCT OF ALL THE PLANTS

AN OCCASIONAL LIGHT FLORAL

HOW TO USE FRAGRANCE IN YOUR OFFICE

CANDLES
Light one at the beginning of work to set the mood, in a bright peppermint or spicy ginger scent.

INCENSE
Use cones or sticks as a timekeeper. Natural rolled incense with resins and florals work best. Avoid heavy sandalwood or Nag Champa so that you stay focused.

REAL MATERIALS
Add in aromatic plants like lavender or rosemary for a pick-me-up, juice a lemon into your glass of water.

While writing this book, we burned about every candle scent imaginable. On the daily, we have candles going on our desks. Some scents are just distracting, and some are invigorating. In my experience, anything with an amber note was too powdery, and left me distracted with a tickled nose.

Occasionally, I'd take scent from real things—walking outside and squeezing a rosemary bush, smelling a piece of cut ginger, squeezing half a lemon into a glass of water. These real, sensory moments provided focus in a different way, as they disconnected my brain and allowed me to focus on the present.

Rosemary is also a fantastic mind stimulant, and has been used as far back as Ancient Greece to improve cognition and memory function. I'm partial to trimming pieces of rosemary off a bush, which is abundant here in Southern California.

There's also a great trick, told to me by our Development Manager Jade: Use incense sticks or cones as a timekeeper. Typically a stick will burn about an hour, and a cone twenty-five minutes or so. Jade will light some incense to set off to do a task, like cleaning for an hour; I would write for the length of the incense burning. Once the incense is done, you can take a break, which is a productivity method (work for forty-five, break for ten) that helps when you're in the middle of a big project.

In terms of scents for focusing during work, I gravitate toward bright and energizing mints, citruses like grapefruit and lemon, and very light florals. I don't want anything too complex (the headier scents might give you a headache when you're focusing very hard), heavy floral (same), or gourmand (unless you want to take a snack break every five minutes).

I don't keep dedicated fragrance in my office space at home because it also doubles as a playroom, so I rely on candles and incense to keep the scent going when I'm working, and I simply take them away when I'm done. Another good bookend for your mind to signify the workday is done.

Outdoors

HOW TO FRAGRANCE THE OUTDOORS

- INCENSE STICKS AND CONES
- BURNABLES, LIKE SAGE AND COPAL
- AMBER CHUNKS OR LOOSE INCENSE MATERIALS, BURNED ON A CHARCOAL DISK
- CITRONELLA CANDLES
- INCENSE (BUG REPELLING) COILS
- CAMPFIRE INCENSE, IF YOU HAVE A FIRE PIT OR CHIMINEA

You may wonder if you need to add fragrance outdoors. I mean, what can compare with the natural fragrance of potted lavender or tomato leaves? Outside is about ambiance through hurricanes holding candles, but incense is also a great option because the smoke won't be overpowering. You can use a light fragrance, similar to the kitchen, to subtly enhance your space. Rosemary, lavender, and basil notes would all be right at home on a patio. When I think of outdoor candles, my mind immediately goes to citronella candles.

I may be the only person who doesn't mind the smell of citronella. I find it clean with an industrial body, and it has notes of bergamot, lemon, and geranium. Hey, fun fact: You can't legally sell or market a citronella candle for bug repelling purposes unless it has actual citronella oil in it. A lot of citronella candles on the market are just citronella *fragrance.*

Back in the day, P.F. used to sell a citronella candle. When we showed it at markets, especially in San Francisco and Los Angeles, people would mosey on up to our Smell Bar, take a whiff, and shout, "This smells like bug spray!" Yes, that was the point. Basically, this is to say, citronella candles are great in some regions (like Austin, Texas, where we started our business), and not so great in others (like dry California that doesn't have a legion of bugs around you constantly).

Another great option for outdoor fragrance that has a dual purpose of bug repelling are those bug-repelling coils you burn on your patio. Growing up, my parents burned those coils on our deck when we'd eat outdoors in the humid Virginia summers. They also remind me of a time when we went to Playa del Carmen in Mexico and we stayed at the excellent Hotel la Semilla, a simple and rustic hotel owned by a sweet couple. Tucked off the main party drag, the hotel is charming and chic, with flea-market vintage furniture, fresh fruit in the morning, and a jungle-inspired back patio. Every room had an incense coil to deal with jungle mosquitos. When we were checking out, one of the owners was burning copal incense sticks, and I have since become obsessed with their spicy and resinous fragrance, even releasing a candle (now discontinued) that captured the body of that scent.

Incense in general is great for the outdoors, as it's atmospheric but the smoke doesn't get trapped in the same way. We're also fond of campfire incense if you happen to have a fire pit—check out the DIY later in this book.

How to Get a Really, Really Good-Smelling House

We've gone room by room and talked about what type of scents and applications work where. But how do you combine it all together? How do you have the type of house that someone walks into and says, "Wow, it smells good in here!" without people getting a headache? When you start to make scented products in your own home, you'll be halfway there. When we are making especially strong scents at our warehouse, you can actually smell them from outside the building. Hopefully our neighbors love us!

For a lot of people, scent can be overwhelming. Tom's sister, and Tom himself, actually have a fragrance sensitivity. People never wear perfume over to her house, and I stopped wearing perfume myself for years when Tom and I first started dating. But Tammy, Tom's sister, happily uses most of our products. What's the catch?

The key is always with dose. We lightly scent our candles and use a low fragrance load so that they aren't overwhelming, meant instead for everyday use. So start by layering. Like a good personal style, the beauty is in the layers.

Start with your base—for me, that's diffusers. I keep one in almost every room as a building block for our home's fragrance. Then, layer candles into each room. Container candles and molded pillars in the living room, beeswax tapers in the dining room, my "rich hippy" smells in the bathroom.

On the weekends, or work-from-home days, or anytime before a guest comes over, we light an incense stick or cone. We do this mainly when we're present or need a dose of fragrance so we can really enjoy the atmosphere it creates.

When our rooms are musty, we spritz room and linen sprays on all the fabrics like sheets and pillows. We do this about once a week.

I also try to open the windows as much as possible, especially if we're burning incense (it's good to release some of that particulate matter).

Lastly, we are selective about the type of fragrances we incorporate into products we don't even make ourselves. Our hand soap is always a lavender variety (Dr. Bronner's)—same with our detergent (Seventh Generation). We don't use scented trash bags (although you *can* spray some room spray in there—hot tip).

Overall, it's helpful to understand that fragrance is in everything, and can say something about you and create an experience for yourself when you are selective about this—whether by purposefully layering fragrance types or looking at the notes like you would a recipe ingredient list.

CALIFORNIA
ELEANOR MAIDMENT

R. R. MARTIN
A DANCE WITH
Bantam

JOHN MUIR
SIERRA
Wabi-Sabi for Artists, Designers, Poets & Philosophers
You Can Train Your Cat
SIMON AND SCHUSTER
CHARLES BUKOWSKI
WOMEN
Burning In Water Drowning In Flame
Charles Bukowski
FACTOTUM
Charles Bukowski
Black Sparrow Press
CAMERON
the ARTIST'S WAY

PARIS
BY DESIGN
Jorgensen
THE NEW BOHEMIANS
HANDBOOK
JULIE CARLSON
Remodelista
CREATIVE SPACES

B.F. Skinner • BEYOND FREEDOM & DIGNITY
THE GIRL WHO KICKED THE HORNET'S NEST
STIEG LARSSON
ECONOMETRIC MODELS AND METHODS
WILEY

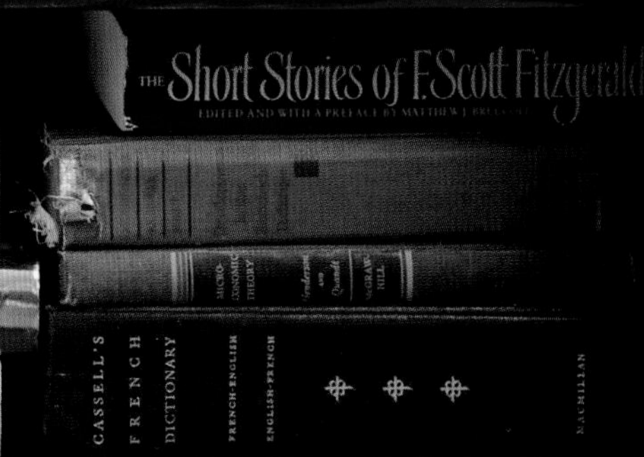
THE Short Stories of F. Scott Fitzgerald
CASSELL'S
FRENCH
DICTIONARY
FRENCH-ENGLISH
ENGLISH-FRENCH

FRAGRANCE

by

ENERGY

Look, sometimes, it's all about mood. I feel you! Obviously, everyone knows lavender is calming, but what about scents that perk you up? Or can get your wheels turning for some creative inspiration?

Scent has a powerful effect on memory and mood. The former comes in the form of the "Proust Effect," named after Marcel Proust, who described the intense childhood memory surge caused by dipping a madeleine cookie in tea. (Yum.) Scientists are still trying to figure out the links between fragrance and your brain, but the theory is that it's related to the proximity of your olfactory gland and the amygdala/hippocampus, which are in control of memory, emotions, and motivation. Scent could help in motivation, processing depression and anxiety, learning, and palliative care.

Further, fragrance and scent communicate without words (part of the reason it is so damn hard to talk about scent!). Scent can tell you something's wrong (if there's a fire or gas leak) before you see the visual cues for the same problem. There's a reason so many scents trigger childhood memories, specifically—this was a time before our language was fully developed, and we relied on our senses to communicate with the world.

Bad smells, or malodors (I imagine these like the Malfoys in Harry Potter, sweeping in with their bad smells like a black cloud), can have adverse effects on emotional, cognitive, and social situations. Scenting the air and removing malodors can improve your standard of living.

All this to say, scent is not only a wonderful psychological mood tool but a therapeutic tool you can use to understand and process your emotions. What I lay out here—the effect of different notes on mood—may change depending on the individual and how those molecules link with their emotions.

A funny example before we get started: Amber & Moss, one of our bestselling fragrances, is one I cannot stand. It's based on a traditional chypre accord, meant to smell like the forest floor—warm but a little damp and green. The reason I don't like it? Back in 2014, Tom was pouring this scent in candles and spilled some raw fragrance oil on his shirt. Just hours after that incident, we piled into a car and drove two hours to the mountains. I never had anything against Amber & Moss before this, but the intensity of the undiluted fragrance oil, and the subsequent presence of that undiluted oil in our tent all night (where I wanted to be smelling trees and campfire) made me dislike this fragrance. That's just the way that memories can influence scent perception. (But, I can see that I'm in the wrong; this scent continues to be our number-two bestselling scent number-one internationally and people find it relaxing and soothing.)

Epiphone
SMELL YA LATER
RECORDS
TAIT
PERPETUAL CALENDAR
W
MAR
17
DSL
40C

Scents to . . .

BRING FOCUS

LEMON

GRAPEFRUIT

PEPPERMINT

ROSEMARY

CINNAMON

TERPENE

JUNIPER

GINGER

Scents that improve concentration and boost your mood tend to be bright and piquant. For a lush effect, pair a bright citrus or mint note with something green or terpenic. This will soften any highly aromatic notes that make your eyes water. Rosemary is known to improve cognition; the herbal and minty mix is a mood booster too. If I'm working from home, and need a break, I'll take a walk into our backyard and squeeze the rosemary bushes a little, releasing their fragrant oils.

Scents to . . .

CHILL YOU OUT

LAVENDER

BERGAMOT

JASMINE

OZONIC, LIKE SEA SALT AIR

MARINE, LIKE COLD WATER

My "happy place" is a spot I call "feet in the Big Sur River." Almost seven years ago, Tom and I took a trip to Big Sur and stayed at a micro cabin with fluffy bathrobes and individual fire pits. During the day, we dragged some Adirondack chairs down to the Big Sur River, drank beers, and put our feet in the cold water. I remember reading (the iconic) Martha Stewart's business book and forcing myself to chill out. It was the first time we'd ever left someone else in charge of the business, even if it was a couple days. When I need to take a real chill pill, I go there in my mind's eye, and remember the river and also the hike we did, through the redwoods and along a sunny ridge through some sun-baked sage. Those are the types of scents—cold-water marine, sea-salt air ozonic, dry white sage, lavender—that help chill me out.

Scents to . . .

GROUND YOU

SANDALWOOD

AMBER

MYRRH

VETIVER

BERGAMOT

Neutral, grounding scents are great for connecting yourself in the present moment. Imagine what a yoga studio would use. It may be a little crunchy, sandalwood, or patchouli, or perhaps a bit more high end like sage or votives.

Whenever we are smelling dozens of fragrances in a row, there will come a moment when you need to "clear your nose." Lots of people smell coffee beans as a scent cleanser, but in actuality, resetting your nose on yourself—the crook of your elbow or the inside of your shirt—is better. So perhaps grounding scents are very personal scents that are intimately you; only you will know what those are.

Scents to . . .

MAKE YOU SMILE

LAVENDER
WARM VANILLA
GARDENIA
BERRIES

When I'm thinking of happy scents, my mind always goes to florals. Florals have been shown to be a mood booster, and their odor has a lot to do with that. Florals like gardenia and tuberose, lavender, warm vanilla, and berries are all mood boosters to me. For the unexpected, you can try a berry note. Strawberry is such a happy scent to me; I associate it with my daughter (who eats strawberries by the pint), my mom (who told me my birthmark—a streak of red in my hair—was because of strawberries), and this bathroom at a bar in Brooklyn I used to go to in my early twenties. A bar bathroom probably wouldn't be the first thing that makes you "happy," but they would always burn a strawberries-and-cream candle in there, and although my early twenties were a time of angsty partying, they were also a time of freedom.

Scents to . . .

COMFORT YOU

AMBER

WOODS, LIKE PIÑON OR SPRUCE

GOURMANDS, LIKE VANILLA OR FIG

OAKMOSS

I'll never forget a customer that told us they used our Amber & Moss candle in palliative care during their father's battle with terminal cancer. They said the scent brought them, and hopefully their father, comfort during that time—and afterward, it reminded them of him. These words meant so much to me, and drove home the powerful psychological impact of scent on your mood. For this reason, I think you should seek out warm scents that feel like a hug (amber), woodsy scents that ground you into nature—which always feels like home to me—or gourmands, of the vanilla or fig variety—there's a comforting simplicity about replicating nurturing food that you eat.

Scents to . . .

REFINE AND POLISH

HEAVY FLORALS, LIKE TUBEROSE

EXOTIC FLORALS, LIKE ORCHID AND LILY OF THE VALLEY

SWEET SPICES

ORANGE BLOSSOM

THYMOL

ALDEHYDES

POWDER

Though you'll catch us in tie-dye T-shirts about half the time, we also certainly appreciate high-quality goods, like nice walnut furniture or real gold jewelry. Fragrance is a great shortcut for feeling a little fancy, no matter how many digits you have in the bank account. The key to luxe scents is layering and texture—complex and heady notes, like exotic florals or heavy sandalwood, make for a rich base. Layer simple spices and a touch of powdery amber, and maybe a bright citrus. Customize and build your fragrance for a one-of-a-kind experience.

HARRISON
NEIL YOUNG WITH CRAZY HORSE
EVERYBODY KNOWS THIS IS NOWHERE
NEIL YOUNG
SUN-C-139
JOHNNY CASH
sings "I Walk The Line"
BOYZ II MEN
Cooleyhighharmony

FRAGRANCE

by

DECOR

Candles, taper holders, incense, and diffusers have an aesthetic benefit beyond fragrance: a decor item. Our classic amber jars took off at a time when many candles were fussy and elaborate, or just not unisex. The simple-verging-on-rustic design felt apothecary, but was cool enough to fit into high-end sneaker boutiques and aesthetically-minded indie shops alike.

With our new line, Sunset, released in 2019, we branched out of our minimalist design and incorporated what we were really feeling at the time: sunsets, horizons, and vintage seventies iconography. This felt like a foray into vessel-as-home-decor.

For candles, I used to scour thrift stores for old glasses and ceramic tureens to pour into. (You'll want to adjust your wick size accordingly.) The same can be done for diffusers—vintage milk bottles and long-necked vases would be great fits.

When it comes to home fragrance as decor, I'm drawn to the dual functionality. Don't get me wrong, I love a few knick-knacks. I still hold onto this vintage lamp with two Siamese cats on it, and Tom and I collect Dodgers bobbleheads. But in my heart, I don't love frivolity in my decor. I want it to work double duty. If I have a solid ceramic doodad, it better also be a vase, or a bookend.

From an aesthetic standpoint, I love that candles are functional art pieces. Candles look especially moody stacked deep in a fireplace, packed by the dozen in vintage taper holders as a centerpiece, or used like a totem—great for molded candles that double as art.

Ceramic incense burners, selenite tea light holders, or a chicly packaged candle all achieve the goal of functionality and fragrance.

When creating our projects for this book, we also became obsessed with creating potpourri. Our house now has more potpourri than a TJ Maxx in 1999. Gone are the days of dried rose petals, however. Potpourri is now inspired by current trends like dried grasses and florals, seed pods, rocks, and shells. Anything you'd collect on a hike can be fragranced and displayed in a beautiful bowl.

A Note on Ambiance

I can't stress enough the importance of scent, candlelight, and incense smoke for creating ambiance. If you think that's frivolous, frankly, you're probably reading the wrong book! (So just go ahead and skip to the DIY secrets, already!)

I always say my biggest design inspiration comes from restaurants—and what's the one thing restaurants (well, good ones) are masters of? Mood. The restaurant at the Freehand in Downtown Los Angeles is one of my favorite spots for this. It's all seventies vibes, with dark wood paneling, retro string art, and vintage (inspired?) glassware. On the tables are sappy, amber-colored candle holders with tea lights inside. I love it when restaurants still use real candlelight!

Side note: You can truly judge a restaurant by its bathroom, and, of course, my favorite restrooms burn P.F.

Candlelight isn't just about the dewy glow it imparts on people lucky enough to sit in its orbit. Candles, or the lighting of them, signify a ritual. Remember our dining room rituals? That came about when I got really into making beeswax tapers, and every night when we started dinner, we'd light one. After dinner, we'd blow it out, signifying the end of the meal.

This ritual, the stopping and starting, helps to create the pathways in your head that say "relax," or "disconnect," or "focus." I use an incense stick from Australia at the start of my bath to say, "OK, we're relaxing now." As soon as I light it, I can step out of my worries a little bit. Bonus: My bathroom always smells a little like that, plus body oils, and a cedarwood and jasmine diffuser. So when I go in there, it feels like a retreat.

Part III

DIY

DIY is the heart of our business—from making products ourselves to building the business without any training. Initially, we were worried about writing a DIY book and "giving away our secrets." But the reality is, making your own products and learning about scent will only make you more appreciative of what a good product looks and (in our case) smells like. In this section, we have fun projects like molded candles, the container candles that put P.F. on the map, and hand-rolled incense (that smells so good, and is so soothing to make).

FRAGRANCE OIL

(OILS USED FOR PROVIDING FRAGRANCE IN CANDLES, DIFFUSERS, AND ROOM SPRAYS)

Fragrance Oil

A catchall term that encompasses oils of natural and synthetic origin.

Essential Oils or Absolutes

Natural oils derived through three processes: cold press, steam distillation, or enfleurage. Essential oils tend to be lighter and fresher compared to fragrances made by other extraction techniques.

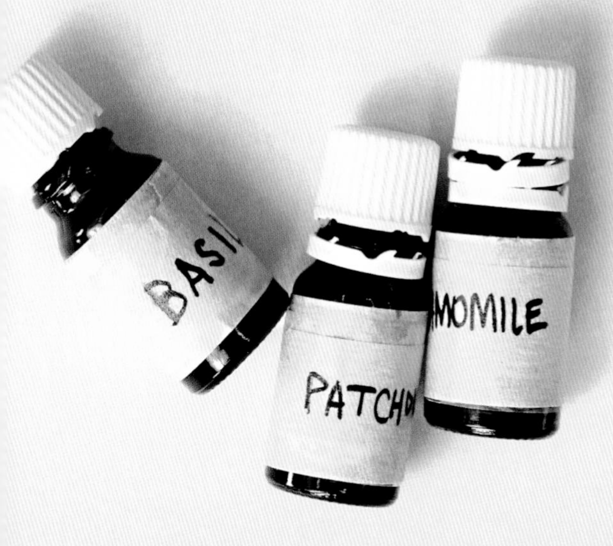

Materials

WAX

Beeswax

Beeswax is a very popular wax—and one of our favorites. The benefits of beeswax are it burns well, it's easy to source, and it has an excellent scent (it smells like rich, warm honey). Sourcing beeswax is the same as it always has been—it's the byproduct of keeping bees. Beeswax also has a very high melting point, so it can be used in pillar and container candles. Beeswax can be pretty pricey, however. In fact, because of colony collapse disorder (where most of the worker bees disappear), the price is increasing all the time. If you are making scented candles, the scent may conflict with the fragrance used in the candle.

Beeswax isn't vegan and there is debate about the treatment of bees in large-scale operations. They are forced to feed on monoculture crops and the honey they live on is replaced with low-nutritional high fructose corn syrup. Even though large operations have a carbon output, because there is no need to clear cut forest to plant crops and the product requires less refining, many view beeswax as a superior ecological choice for candle making.

Vegetable Waxes

We can now distill waxes from multiple types of plants. Some popular vegetable waxes are soy, coconut, palm, and apricot. To make the wax, oils are extracted from the plants and then the melting point is raised to turn these oils into solids through a process called hydrogenation. Each wax has different properties with their own pluses and minuses. Vegetable waxes tend to be more eco-friendly because they are more carbon-neutral than paraffin waxes. Some are better than others, however. For example, we never use palm wax, because harvesting the palm is particularly destructive to natural environments. Across the board, some drawbacks are that vegetable waxes tend to be harvested from monoculture farms, or farms may use GMOs, which many find objectionable, and the refining process uses heavy chemicals, such as hexane. However, we still feel it's a superior choice to paraffin, which is nonrenewable.

Soy

Soy wax is our bread-and-butter. We've been using it since 2006 to make candles. It was originally popularized as an alternative to paraffin wax. In the United States, soy is domestically grown and an abundant, renewable crop. What we've noticed from our use in the factory is that soy has a slightly acidic vanilla note that complements a lot of the scents we use. Soy is notoriously hard to work with, however, which is why a lot of candle manufacturers blend their soy with paraffin or other additives. Soy doesn't set well, and the scent throw is less than paraffin.

Coconut

Coconut wax has a strong smell of its own—something to consider when blending with fragrance. It also has an extremely low melt point, making it difficult to work with; your candles will start to sweat when the temperature hits the seventies! Coconut/beeswax blend waxes are very popular for this reason. Coconut wax has good scent performance, but a much sootier burn.

Petroleum Waxes

Paraffin is a byproduct of the distillation of petroleum. During industrialization, paraffin became very popular because it was cheap, had a very high melt point, set well, and burned a lot cleaner than most of the alternatives at the time. I think most candles made today are still made out of paraffin because of these points. Paraffin has fallen out of popularity though, because it tends to soot more, while releasing a chemical scent (like burning plastic). It also has a worse carbon footprint, as it both takes carbon from the ground and releases it into the air (unlike a plant wax, that puts carbon back into the ground). Even if vegetables and beeswax are not entirely carbon-neutral, they definitely have a lower carbon footprint than material extracted from the ground. Paraffin waxes are all vegan.

Choose whatever wax makes the most sense for what you are doing. Are you looking to make a low eco-impact, scentless pillar candle? Try beeswax! Are you looking to make candles very economically? Try paraffin! Are you looking to make tropical scented candles? Try coconut! This is where the artistry starts in candle making.

WICKS

Wicks are porous materials that help soak up fuel (wax, kerosene, etc.) and bring it to a flame. They are usually made of braided materials like cotton, though wood wicks have grown in popularity over the last decade. Different braid patterns are used for different purposes. For example, beeswax tapers take a square wick that is engineered to curl while it burns, which can reduce carbon buildup.

Today's wicks have the ability to self trim, which saves us all a lot of hassle, otherwise the wick would stay the same size and as the combustible material lowered, the wick would become larger. Some have metals in them to control the burn. Metal wicks aren't desirable because of metal pollution.

Which wick to use? Cotton wicks are abundant and reliable. Wood wicks give the candle a rustic campfire feel as they crackle while being burned. At our HQ, we use cotton wicks coated in vegetable wax. Follow the guides on the website where you buy your wicks, and then use trial and error. Truthfully, the best way is to test. We run pretty robust tests with three sizes of wicks when we want to set up a new wicking system. For these projects, we've done the guesswork for you.

There are a lot of safety issues to think about here. A candle that burns too hot can be a fire hazard. A candle that burns too cooly and self-extinguishes is only a disappointment, not a danger to anyone.

The thickness of the wick directly correlates to the heat of the flame: the thicker, the hotter; and the thinner, the cooler. For the home candle maker, wicks are available two ways:

Shim

Precut wicks with a metal shim at the bottom. The metal shim keeps the wick from burning too low into a candle. These are great for container candles.

No Shim

Spooled wicks for molded candles and tapers.

DYES

For candle making purposes, you can find liquid dyes or chip dyes. We've always used the little chips because it just seems easier. Dyes are another area that requires quite a bit of experimentation to get the color right.

CHARCOAL

Charcoal is a very important material in incense making. In the most simple form, charcoal is what you burn other materials on. When making incense, charcoal can be a good base to make materials more combustible. There are instant-light types and plain charcoal, and there are many opinions about which one is best. Instant-light charcoals have little divots on the top to keep the material in place and light much quicker, and they're what you'll use when burning loose incense. They do contain additives and burn very hot, which doesn't make them suitable for combustible incense bases.

PLANT MATERIALS

Our incense and potpourri projects call for "raw plant materials"—what that means is ground or loose flowers and leaves from plants and trees like lavender, rosemary, cedar, and juniper. For potpourri or non-combustible loose incense, you can use plant materials whole or chopped up. For combustible or stick incense, you're going to want to purchase it powdered or pre-ground.

RESINS, GUMS

Incense also calls for resins and gums, like frankincense, myrrh, and copal. These can be ordered in a chunky, almost crystal form and then powdered, or you may find them pre-ground (which is highly recommended). Gums and resins can be frozen before you powder them, which makes them easier to crush. Although these might be a little more foreign than plant materials in incense making, it's highly recommended to incorporate them into your ingredients, as they give scents a round and robust feeling.

MAKKO POWDER

Makko powder is our choice of combustible material when making hand-rolled incense. It's less powdery than charcoal.

FIXATIVES

Orris root is a fixative that we use for our potpourri recipes, as it helps the scent stay for longer. You won't be frequently using fixatives, however.

1
2
3
4
5
6

Equipment

SPICE GRINDER

Some of the materials you are going to get for incense, like sweetgrass or resins, are going to be in a pretty solid form. Grinding them down allows for them to be mixed more evenly. You can use an old coffee grinder, or anything similar that will not be used again for food. Mortar and pestles work well too, if you're into the old-fashioned way of doing things. I use an old molcajete that was given to me for making guacamole.

APRON • COAT • OVERALLS

Working with fragrance can be messy. Fragrance and wax spills can stain clothing because they are oily and can soak through to your skin, which can cause irritation. Having a nice set of work clothing or an apron can prevent the need to go shower down after a big spill.

STOVETOP

If you don't want to spend a ton of money on candle making equipment, don't. Whatever is cheap and fits your needs is perfect. We made candles on a stovetop for the first six years of our business using a double-boiler system. If you have a kitchen, you have the equipment to melt the wax. You can also use an electric hot plate.

EYE COVERING

When working with liquid materials, it is recommended that you have some sort of eye covering. They look nerdy, but considering they could keep you from going blind, it is worth it.

MELTER

Depending on the amount of candle making you are doing, a melter can be a big time saver. These usually start in the $1,000 range, so it is an investment. A lot of melters operate like open-top water heaters. Two types of melters are readily available: water jackets and direct heat. Water jackets have heating elements within a jacket of water. They heat the water that then heats the contents inside. Direct heat has the heating elements embedded inside the walls directly heating the material within. Water jackets tend to heat more evenly and quickly because some have multiple heating elements. They are more expensive and take up more power. Direct heat is cheaper and uses a more reasonable amount of power. They do take longer to heat. Our first melter was a fifty-pound water jacket. It took so much power that when we moved into our first studio, if we ran anything on the same breaker it would cut the power off. Every melter we've used since that time has been a direct heat melter. If you have a crafty spirit, melters can be used for other projects, like soap. If you're looking to turn a hobby into a business, this is a good investment, but not necessary.

POTS

An old pot full of water with a pitcher of wax inserted inside is a great double-boiler system. You fill a pot full of water, put a pitcher in that water, then boil it. You could use your current set of pots, but be warned, candle making will ruin them pretty quickly. (Just ask Kristen's mom, who had no idea she was donating her nice pots for eternity to the cause of her daughter's business.) Wax and fragrance will get into that boiling water at some point. The minerals in your water stay as you boil the water out. So if you've invested in nice cookery, just go buy some cheap metal pots to boil water in.

GLASS BOTTLES

It's helpful to have varying sizes of glass bottles to keep all your liquid fragrance in. We use a range from 8 to 32 oz. At our studio, we use amber Boston rounds, which have a tapered top that makes pouring easier and spillage less of an issue. We like the amber color because some fragrances can oxidize when exposed to the sun. The amber helps to keep the sun out, thus preserving the state of your materials. Ever think about why most beer bottles are amber? This is why! Also, amber jars are kind of our look.

/ 1

WOODEN SPOONS

Pitcher mixing can be done with a standard cooking spoon. When we were mixing in melters, we would use an oar to do the mixing. It had a real witch's kettle vibe.

/ 2

THERMOMETER

Two types of thermometers work here: candy thermometers or infrared thermometers. Candy thermometers are giant glass thermometers you place in material to measure the temperature. Infrared thermometers are little electric guns that measure the heat by reading the infrared energy of a material. Candy thermometers tend to be a little cheaper and more accurate since they sit in the material. Infrared can be a little more expensive and can be incorrect if it is catching the reading of something else in between, but it's the only type of thermometer we use in our studio.

/ 3

TAPE

For projects here, the tape is primarily used to keep the wick in place in the molded candles.

/ 4

SCENT STRIPS

These are little paper strips used for dipping in oils to sample fragrance. This is the building tool for creating a fragrance.

/ 5

GLOVES

Thick rubber gloves are going to give you the most protection, but offer limited mobility. I always use a thick latex or nitrile glove that gives you a mixture of protection and dexterity (the kind tattooers use).

/ 6

HOT GLUE GUN

You'll be using this for gluing wicks to the bottom of vessels. There's no need to buy anything fancy, the cheapest glue gun works. Just be careful as the metal tips get hot, as you can burn yourself if you aren't paying attention.

/ 7

FUNNELS

You'll need funnels of all sizes for pouring liquid from one container into another. We recommend metal ones because some of the fragrance oils can be caustic to plastics, so the metal will last a lot longer.

/ 8

POURING PITCHER

You will need a metal pitcher that you mix your candle making materials into and pour out of. Pouring pitchers are such a utilitarian tool for us. You can mix, weigh, melt, store spoons, shovel wax. If you're going to be making more than one candle, it's worth the ten-dollar investment.

/ 9

PIPETTES

These are used when mixing liquid fragrance together. With a pipette, you can gather and dispense small amounts of liquid material, allowing you to accurately collect what you need.

/ 10

SCALE

Our first scale was a cheap postal scale that worked double duty: as a package scale and a material scale. We've bought many over the years. Just make sure you get one big enough that it will hold your materials. Once, Kristen ordered us a scale for development, only to receive a tiny pocket drug scale. That didn't work well for us at all.

/ 11

PUTTY KNIFE

You'll use a putty knife to scrape the rough edges of any of the molded candles, or scrap wax off surfaces.

Building a Fragrance

Building fragrances is one of the most rewarding parts when learning about home fragrance—but it can definitely be intimidating when you first start. Here are some formulas for fragrance oil blends that you can use as the fragrance oil component in any of our candle or diffuser projects.

For our sake, we'll say that we're blending four ounces for each of these oils. That's enough to make two candles and a room spray or diffuser. Once you smell these blends, you can adjust the percentage of the different types of oils to your liking.

When you're initially smelling these notes, dip a blotter strip (or a piece of paper, if you don't have those) into the oil, just a bit. Make sure you label your blotter strip. Then smell the strips altogether, pulling one back away from your nose to create the effect of a lesser percentage of oil in the blend.

Many of the oils you'll find commercially available on the market are pre-blended accords, so you'll have to dig within the manufacturer's description to find a simple scent like, for example, a basic sea salt. It's fine to use accords in place of these notes, and, in fact, will give your oil individuality and character.

COASTAL REDWOOD FOREST

EVEN THOUGH WE LIVE in Los Angeles, smell-wise, Northern California and its redwood forests have our hearts. We spend a lot of time in the Bay Area at markets, and now at our shop in North Beach. Our favorite overnight stay is camping at Mount Tamalapais, which smells damp, misty, and woody—a little crunchy (in the lifestyle sense and the leaves sense).

---

2 OZ REDWOOD FRAGRANCE OIL

½ OZ SEA SALT FRAGRANCE OIL

1½ OZ EUCALYPTUS FRAGRANCE OIL

DELUXE

THESE THREE NOTES are each dynamic superstars in our fragrance arsenal. Combined together, they are a sophisticated and chic unisex fragrance. The rich notes create a sense of mystery and body. I've kept the bergamot essential oil to a minimum here, to keep costs in mind, but feel free to adjust as you like.

---

1 OZ SANDALWOOD FRAGRANCE OIL

2 OZ PETITGRAIN FRAGRANCE OIL

1 OZ BERGAMOT ESSENTIAL OIL

EASTERN SIERRA

THE EASTERN SIERRAS in California is one of our favorite places. In 2015, before the birth of our daughter, we took a backpacking trip there through the Ansel Adams Wilderness. We climbed to a very high Alpine lake fed by a glacier, took pictures, and read books for a couple days while sleeping in a very tiny tent. The Eastern Sierras have a cleansing effect on both your mind and senses, cleaing you out like a natural VapoRub. Our favorite types of fragrances are those inspired by nature. Look for fragrance oils that are blended with essential oils, or pick one of the notes below and use it as a straight essential oil.

---

2 OZ JUNIPER FRAGRANCE OIL

1½ OZ WHITE SAGE FRAGRANCE OIL

½ OZ VETIVER ESSENTIAL OIL

APPLE PINE

THE COMBINATION OF AN APPLE scent and a coniferous tree here is inspired by a legendary mistake that happened on our production line a couple years ago called "Apple Spruce." A candle pourer mistakenly added our Spruce scent into the melter along with Apple Picking. At first we were annoyed—this was a huge gaffe in the middle of our holiday season! Then we smelled it, and we were delighted. We created a special label, after debating the merits of Apple Spruce vs. Sprapple Picking as the name, and sold it as a deep cut on our website. People still ask about it to this day.

½ OZ CINNAMON FRAGRANCE OIL

½ OZ CLOVE FRAGRANCE OR ESSENTIAL OIL

1½ OZ APPLE FRAGRANCE OIL

1½ OZ PINE, SPRUCE, OR ANOTHER EVERGREEN FRAGRANCE OIL

VACAY MODE

I WAS NEVER PARTIAL to tropical or coconut scents until we started more regular pilgrimages to the beach after the birth of our daughter. Now, in the middle of winter, a whiff of sunscreen will transport me right back to sea salt in your mouth and sand in your toes (and, well, your everywhere). I created this with a beach vacation in mind. The lime is reminiscent of margaritas, and orange blossom is a key component for that gritty sunscreen feel.

1½ OZ COCONUT FRAGRANCE OIL

½ OZ LIME FRAGRANCE OR ESSENTIAL OIL

2 OZ ORANGE BLOSSOM FRAGRANCE OIL

FRESH LAUNDRY

LAUNDRY IS A SCENT that people go absolutely nuts for. Personally I use lavender scented laundry detergent, but I still like the fresh, dried cotton effect that laundry scents impact. Musk is key here for creating the freshly-dried-towel atmosphere.

---

1 OZ WHITE MUSK FRAGRANCE OIL

2 OZ LAVENDER FRAGRANCE OIL

½ OZ LIME FRAGRANCE OR ESSENTIAL OIL

½ OZ LEMON FRAGRANCE OR ESSENTIAL OIL

CLASSIC CHYPRE

YOU MAY NOT KNOW you love a chypre scent, but I guarantee it's a fragrance profile that's familiar. Chypre (pronounced shee-pra) scents are described as being reminiscent of the forest floor in autumn, with notes of oakmoss, labdanum, patchouli, and bergamot. This scent profile is thought to date back to Roman times. The patchouli is key here for creating that earthy dirt vibe. Amber stands in for the traditional labdanum component, which is harder to find.

1½ OZ OAKMOSS FRAGRANCE OIL

1 OZ PATCHOULI FRAGRANCE OIL

1 OZ AMBER FRAGRANCE OIL

½ OZ BERGAMOT ESSENTIAL OIL

NEW MEXICO

IN LATE FALL OF 2019, we took a road trip across Arizona to New Mexico, visiting Santa Fe and Georgia O'Keeffe's house, driving down two lane highways, and stopping in random spots to take pictures for our 2020 zine.

At night, we built fires in the kiva fireplace of our Airbnb rental with piñon and juniper; during the day, we walked in dormant lavender fields and smelled essential oils. We took in cold air, dry grasses, and a touch of sweetness. We smelled juniper bushes and creosote and red earth. This blend is inspired by those travels, and can show you how you can use your own scent experiences during travel to create a memory-inspired fragrance.

½ OZ SMOKE OR BIRCH TAR FRAGRANCE OIL (IF THIS IS HARD TO FIND, LOOK FOR A CAMPFIRE SCENT)

1½ OZ PIÑON OR PINE FRAGRANCE OIL*

½ OZ JUNIPER FRAGRANCE OIL

1 OZ LAVENDER FRAGRANCE OR ESSENTIAL OIL

½ OZ EUCALYPTUS FRAGRANCE OR ESSENTIAL OIL

**If you can't find a piñon fragrance oil, you should add a touch of vanilla for sweetness.*

Essential Oil Blends

For the room spray project, or if you'd like to work with essential oils for your candles, we put together some essential oil blends here, using a formula where about half of the mixture is mid notes, and the rest split between top and base, with more toward the top. (One thing to note about using essential oils in your candles is you'll have to watch the temperature at which you add these—essential oils have a low flash point, so if you add it while the wax is too hot, all your top notes could burn off before your candle even solidifies.) These would also be great in an essential oil diffuser.

#1

Our 2014 Blend

20% Bergamot
30% Sage
50% Cedar

#2

Garden Blend

25% Basil
25% Lavender
50% Rose

#3

Flower Power Blend

30% Neroli
40% Chamomile
30% Patchouli

OUR FAVORITE
FRAGRANCES

Kristen's Favorite Scents

Woodsmoke

Sunbaked sage

Lavender

Rosemary, crumbled between your hands

Asphalt after the rain

Freshly struck match

Piñon incense

Tom's Favorite Scents

Nag Champa

Anything terpenic (pine or cedar)

Anything ambery

Wintergreen

Angelica seed

Basil

Soil

Vetiver

Galbanum

Garlic

CANDLES

Candle making is pretty simple, once you have all the tools. From 10,000 feet, it's essentially melting wax down, adding fragrance if you'd like, and then pouring that into a wicked vessel or mold (or dipping a wick into wax). But every day we get questions from aspiring candle makers about the specifics. What type of wax is the best? How much fragrance should we add? Which wick should we use? Answering these questions is the hard part.

For years, candles existed in a realm of utility. In a world without electricity, fire was the way people lit their homes at night. Now people use candles for many purposes beyond illumination: aromatherapy, mood, decoration, ritual.

Although we could dedicate an entire book to candle making, we narrowed it down to a few of our favorite candle projects in this chapter. Container and travel candles allow you to focus on the art of the fragrance and vessel. Rolled tapers highlight the materials—sweet-smelling beeswax. Pillar candles experiment with dyes and colors and shapes, serving a role both artful and functional. Candles are endlessly customizable once you get the hang of it.

CONTAINER CANDLES

WHEN IT COMES TO DECORATING, container candles can take all sorts of shapes, sizes, and ambiances. Containers can look like vases, planters, or glasses. For us, container candles were both the start of the P.F. Candle Co. and what got us into everyone's house. Back in 2008, while working at a craft magazine, Kristen started an eco-friendly venture, turning used goods into new homewares. One of her first products was a candle in an old teacup. That wasn't very scalable, so she found a jar that looked vintage, with amber glass, and put a hand-stamped label on it. The scents were just things she liked off the shelf. It was a humble candle that came to define a certain design aesthetic.

For this project, I'm going to create a container candle using soy wax and a fragrance (check our fragrance blend section for inspiration). I've been making so many candles that I feel containers are in my DNA at this point.

This project is for four candles, but once you have the basic formula, you can stretch it out. You could make hundreds of candles using this method (you'd probably want to double to eight at a time, though). For our first major account order (West Elm), I fragranced each batch in the pitcher and could only make one to two hundred a day. It wasn't easy, but I got it done.

To make a project, you'll need to find some three-inch-wide jars. We used a straight-sided jar here, but you can use a Mason jar, or even something you find at the thrift store. If your vessel is wider than three inches, you'll just need to adjust your wick size. Check with the supplier of your wicks for a wicking chart if you size the vessel up or down. These jars should be able to hold about eight ounces each.

—TOM

YIELD

MAKES 4 CANDLES

SUPPLIES

METAL PITCHER

METAL COOKING POT (APPROXIMATELY 3 QT—ENOUGH TO FIT THE METAL PITCHER INSIDE WITH SOME ROOM)

STOVE

SCALE

INFRARED (IR) OR CANDY THERMOMETER

TIMER

4 CLOTHESPINS

SCISSORS

MATERIALS

30 OZ OF WAX

WICK ADHESIVE (HOT GLUE OR WICK STICKER)

4 ECO-8 OR LX-8 WICKS

FOUR 3-INCH-WIDE CONTAINERS (SOMETHING AROUND 8 TO 9 OZ)

2 OZ FRAGRANCE

1. First, you'll want to get your wax melting. For this project, you can use a double-boiler system by putting the wax into a metal pitcher and placing that pitcher into a pot full of water on the stove. Turn the stove to high and bring the water to a boil.

2. While the wax is melting, prep your jars for pouring. Take wick stickers, or a dab of hot glue, and place it on the bottom of the metal part, or shim, of the wick. Adhere the wick to the center of the vessel. You can use a shortened paper straw or a cut-off pen tube to help you get a firm adhesion.

3. Use a thermometer to check the temperature of the wax. We are looking to get the wax to about 180°F. This temperature opens up the wax, and allows for the fragrance to easily diffuse, but isn't too hot to cause the fragrance to evaporate out of the pitcher. Once it's reached 180°F, add the fragrance, using a scale to add a precise amount.

4. Set a timer for 2 minutes and stir that wax. The stirring diffuses the fragrance throughout the wax.

5. Allow the wax in the pitcher to cool to about 135° to 140°F. A lot of variables go into the pouring temperature, but we've generally found that 20°F over melt point is a good place to start when figuring out your pour temperature.

6. Pour the wax into the container. Something to consider here is the speed of the pour. The more even and slow you can pour, the less air that will get caught in the wax, which means your candle will set better. You want to imagine you are pouring a beer and are trying not to get any head.

7. Use a clothespin to hold the wick upright in the vessel, clipping it in place so that the wick is perfectly centered. The wick should be very taut so that it doesn't have any curve—this could allow the wick to extinguish or move, which could heat up the glass too rapidly.

8. Let candles cool for 24 hours. The key for a smooth set here is even cooling. Candles need to cool from the bottom to the top. If the top cools first, heat will get trapped in the center and the candle will not set properly. To allow for even cooling, pour your candles in a temperature-controlled room away from drafts, and space the candles about a hand's width apart.

9. When candles don't set smoothly, there may not be an issue when it is for personal use. As long as there are not large craters or cracks near the wick, the wax may smooth out after the first burn. If the candle is for someone else, and you want the candle to look one hundred percent, you can refinish with a heat gun or hair dryer, or you can reserve a small amount of wax and pour a thin smooth coat on top.

10. Trim your wick to ¼ inch before burning.

TRAVEL CANDLES

TRAVELING CAN BE A VERY FUN ADDITION TO YOUR LIFE—getting to experience parts of the world, learning about different cultures (and their scents) through food, landscape, shopping, and architecture. On the one hand, travel is very satisfying and good for us, but on the other hand, it can be lonely. If you are traveling by yourself, you could be missing your family. If you are traveling with your family, but deeply immersed in an unfamiliar culture, you could miss home. Scent can be a quick and easy way to make your new surroundings feel very familiar.

When we first started the company, our dream was to be a mobile candle shop, roaming through the United States selling candles out of the back of a Volkswagen bus. That dream never fully materialized for us, but we did a fair amount of travel for the business: heading up to San Francisco four times a year; flying to Washington, D.C., and New York for craft fairs; visiting our European Union distributors at their headquarters. The trips would all start exciting, but soon we would miss our friends and family. Luckily, we always had fragranced products with us to help remind us of home.

The first time we ever made travel candles was as guest favors for our friend's wedding. Great gift, right? Small enough to carry home, with a long-lasting use time. They were these Teakwood & Tobacco soy candles in gold tins. We placed a sticker on top with the wedding information to personalize them. At the time, I was always focused on trying to really streamline production and make as much product as possible that I moved really quickly. That didn't always work out well, like in this instance where I ended up dropping all the travel tins on the ground. The tins got all banged up, and I had to pick through what we had in order to salvage enough of them to deliver the product. This is a cautionary tale that the tins can be tricky to work with. They are great for personalizing, though, since you can find different colors online, or go the classic silver route and affix a simple label.

—TOM

YIELD

MAKES 5 CANDLES

SUPPLIES

METAL PITCHER

METAL COOKING POT (APPROXIMATELY 3 QT—ENOUGH TO FIT THE METAL PITCHER INSIDE WITH SOME ROOM)

STOVE

SCALE

IR OR CANDY THERMOMETER

TIMER

5 CLOTHESPINS

SCISSORS

MATERIALS

13½ OZ OF WAX

WICK ADHESIVE (HOT GLUE OR WICK STICKER)

ECO-6 OR ECO-8 WICK

FIVE 4-OZ TRAVEL CONTAINERS

1 OZ FRAGRANCE

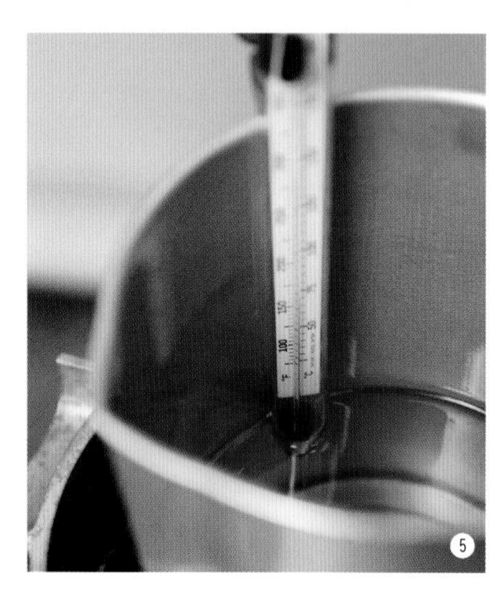

1–7. Follow the instructions for Container Candles (see page 113).

7

ROLLED TAPER BEESWAX CANDLES

TAPERS ARE MADE WITH A HIGH-MELTING POINT WAX in order to hold their shape without a vessel. There are a few ways to make a taper candle. The first is the classic dip method, where you dip your candles in a vat or pot of wax. After that, you dip the candle into a pot of water to cool. Each dip adds a coat on top of the previous. You repeat the dipping until you have a handmade and organically shaped taper candle. You could also pour the wax into a taper mold. This is an easier process with a more accurate product, but the molds take longer to cool. Finally, you can make rolled tapers, which involves rolling sheets of beeswax around a wick. Beeswax is one of my favorite waxes to work with, and this project is so accessible because all you need are sheets of wax and wicks—no pots, molds, or fragrance to mess with.

When I was a kid, my mom would mount colored versions of these on stakes and put them in the garden. Did she ever light them? Can't remember, but the blue and purple beeswax looked cute. Personally, I prefer the look of natural beeswax.

Beeswax has a natural honeyed scent, warm and comforting but mellow enough that it can accompany dinners without being distracting. These are so quick to make, that if you're trying to start a new habit of candlelit dinners, you can quickly roll them in the time it takes to set the table.

Roll these tapers up, wrap them in some twine, and voila—you've got yourself a great housewarming gift! I saw this tip on Erin Boyle's blog, *Reading My Tea Leaves*. She's all about sustainability, and these tapers are an excellent zero-waste gift. Once they burn down, they're gone, without leaving any waste behind.

—KRISTEN

YIELD

MAKES 2 CANDLES, BUT THE PROJECT IS EASILY ADAPTABLE TO MAKE MORE

MATERIALS & SUPPLIES

SHEETS OF BEESWAX (THEY TYPICALLY MEASURE 8" X 16")

SQUARE BRAID, SIZE-1 WICK

SCISSORS

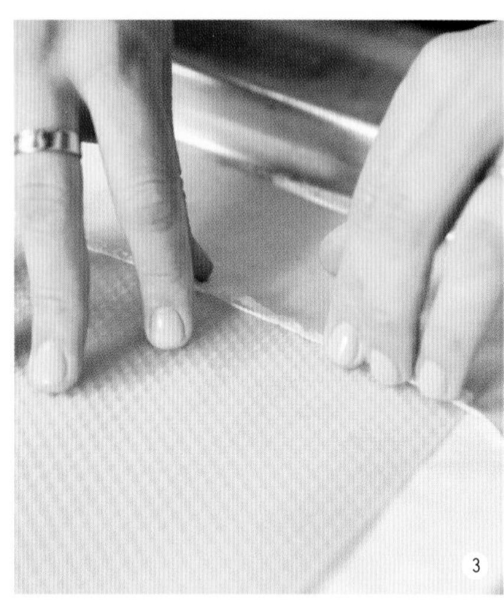

1. Taking an 8" x 16" sheet of beeswax, cut it half short ways, so that you are left with two squares of beeswax.

2. Place your wick along the bottom edge, about ½ cm up from the end of the wax sheet.

3. With your fingers, gently squish the beeswax sheet over the length of the wick. Wrap this part tightly, so that it holds your wick in place at the core.

4. Once you've got the first wax folded over the wick, keep rolling. You'll want to keep these rolls tight to eliminate air pockets and make your candle burn better and longer.

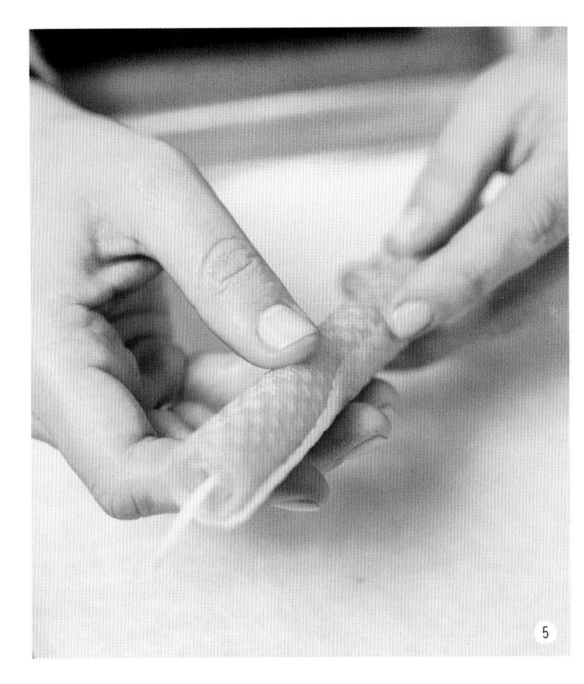

5. When you come to the edge of the sheet, simply compress the edges lightly with your finger to make it stick. Beeswax is naturally tacky, so it shouldn't be a problem.

6. Trim your wick to ½ inch on one side (the top), and flush to the candle on the other (the bottom).

7. If you'd like to make a bigger size and keep on rollin', switch your wick to a size 3 or 4.

MOLDED CANDLE

BACK WHEN I WAS FIRST STARTING POMMES FRITES, one of the things I was known for were owl candles. Each one was molded with a vintage mold I scored on eBay. At the time, most of the sales were made around Christmas, so the best-selling scents were always holiday gourmand: Pumpkin, Gingerbread, Apple Cider. I have a lot of nostalgia for that time. Those scents are so warm. The act of making those candles literally warmed up our house. Because of this memory, when we light any of those candles, I get such a warm fuzzy feeling.

Molded candles are a very fun way to decorate your house. There are thousands of molds available on the internet for you to try, allowing endless customization. There are molds shaped like pyramids, squares, and hexagons, molds shaped like animals and anchors and unicorns. There are plastic molds and aluminum molds and silicone molds. Needless to say, there are a lot of options.

The cool thing about plastic molds is that you may actually find some vintage ones, like the owls that were among the first P.F. candles. Plastic degrades over time, however. Aluminum molds will last you forever, but the craziest shape you'll find is a pyramid or hexagon. Silicone molds are probably my favorites to work with because they self-seal and have fewer leaks. Plus, if you're very crafty, you can create your own molds, using silicone mold rubber, into whatever shape you want.

—KRISTEN

Determining how much wax you need: Each custom molded candle is unique! This is truly trial and error—you'll probably want at least one if not two pounds of wax to start. Any leftover can harden inside the pitcher and you can re-melt it for future projects. It's always better to have too much instead of too little—that way you don't have to do math to figure out wax or fragrance formulations.

YIELD

MAKES 1 CANDLE

SUPPLIES

- METAL PITCHER
- CANDLE MOLD
- TAPE, IF USING A PLASTIC MOLD
- NEEDLE, IF USING A SILICONE MOLD (TO HELP THREAD THE WICK)
- CLOTHESPIN OR SKEWER, TO HOLD THE WICK IN PLACE
- MOLD SEALER PUTTY (FOR PLASTIC MOLDS)
- SCALE
- IR OR CANDY THERMOMETER
- STOVE
- TIMER
- SKEWER
- PUTTY KNIFE FOR SMOOTHING EDGES OF SEAMS

MATERIALS

- WAX (APPROXIMATELY 28 OZ FOR THE HORSE SHAPE)
- MOLD
- WICK (WE USED LX-20)
- FRAGRANCE
- DYE

1. Start melting your wax in your pouring pitcher, set inside a double-boiler system of a pot filled with water. While your wax is melting, prepare your mold: Calculate the wick size you will need using an online wick chart or consulting your wick manufacturer.

   a. If using a plastic mold, tape the wick in place at the top of one side of the mold. Secure at the bottom as well. If using a silicone mold, I find it's easiest to use a large needle (used for knitting or darning) to push the wick through the silicone. Secure at the top using a clothespin or skewer. If using a plastic mold, you will place the two sides together and then secure in a mold stand. You'll want to make sure this union is tight so that no wax leaks out; you can use mold sealer putty to secure all the edges.

2. Once your wax is fully melted and has reached 180°F, add your fragrance and dye. Fragrance is best added using a scale so that you don't over fragrance.

3. When you're working with dye, you'll need to stir vigorously in order to fully disperse all the dye. For fragrance, use 1 ounce for each pound of wax.

4. Once the dye and fragrance is dissolved (after a couple minutes of stirring), remove the pouring pitcher from heat and allow it to cool to about 140°F.

5. The first time I pour into a mold, I pour a small amount to test the mold for leaks. If no wax comes running out, I'll continue pouring until the wax reaches the top of the mold (or if using an adjustable size, until it reaches the desired height).

6. Next, you'll need to create relief holes. These prevent air bubbles from occurring along the wick—which could result in an improper burn—and also create a smooth bottom. It will take about 45 to 60 minutes, depending on the size of your mold; the wax should have solidified enough that when poked with a skewer or stick, the wax is solid but still malleable. Poke a few relief holes on the top of the wax.

7. Reheat the remaining wax in the pouring pitcher, then pour melted wax over the relief holes.

8. Allow the candle to cool for at least 6 hours before removing from mold. Carefully remove, especially if there are any odd shapes on the mold like animal ears or flower buds. These could crack at this point.

9. If using a two-piece or a seamed mold, you can use a putty knife to smooth the edges by running it along the wax seams. If the bottom of the candle is not level, you can also use the putty knife to smooth it out at this point (making sure you face the putty knife away from your body).

10. Trim your bottom wick flush with the candle. Trim your top wick to ¼ inch.

TRIPLE-STRIPE CANDLE

WORKING WITH MULTI-COLORED CANDLES IS REALLY FUN. I used to make two-tone candles as gifts for my family in high school. I've updated this from the frosty, paraffin affairs of my youth to chic striped and scented creations.

What I love about these candles is how they double as decor. The bonus is that you can build and customize for different aesthetics. For my sister, I might make a simple white and teal stripe. For our shop manager, I'd bust out the pitchers and pour a neutral rainbow of brown, tan, and cream.

For this project, you'll need a little bit more in the setup department. If you don't have multiple pouring pitchers, a Pyrex cup will do—just be careful if you're using a double-boiler system. You can melt wax in your metal pitcher and pour off into a Pyrex cup to tint and scent.

The fun part of this project is the combination of color and fragrance. Yellow + green would be a great opportunity to create an accord with sunny citrus and a leafy green scent. These are especially fun for the holidays, with one layer of Mistletoe and one layer of Vanilla. Keep your individual fragrances pretty simple, with single notes or simple fragrance accords so that they can harmonize when you melt the layers together.

For this project, we used one dye chip to gradually darken the wax, creating an ombre effect. If you'd like to make different colors, you'll need multiple pitchers, or you could use multiple Pyrex cups.

—KRISTEN

YIELD

MAKES 1 CANDLE

SUPPLIES

MOLD SEALER OR SILICONE MOLD PLUG

BAMBOO SKEWER OR STICK

POURING PITCHERS (ONE FOR EACH COLOR, IF YOU'RE DOING DIFFERENT COLORS)

METAL COOKING POT (APPROXIMATELY 3 QT—ENOUGH TO FIT THE METAL PITCHER INSIDE WITH SOME ROOM)

3-INCH METAL PILLAR CANDLE MOLD (YOU'LL WANT ONE THAT IS AT LEAST 8 INCHES IN HEIGHT)

CLOTHESPIN OR ANOTHER SKEWER, TO HOLD THE WICK IN PLACE

MATERIALS

1 LB PILLAR SOY WAX

WICK (WE USED LX-22)

FRAGRANCE

DYE CHIPS

A note on dye and soy wax:
I've always worked with dye chips for my projects. When you use these in soy wax, they tend to be pretty light, almost pastel. Personally, I'm not a vibrant color person, so this never has bothered me, but it's worth noting as a flaw with chip dyes. If you want more vibrant colors, use liquid dyes. With dye chips, I've given a rough idea on how to create the colors pictured, but you'll really want to test by dropping a bit of wax onto a piece of paper to see what the color looks like before pouring.

1. First, add wax to your pouring pitcher. While your wax is melting, you'll want to prepare your mold. String your wick through hole the bottom of the mold. Secure your wick hole with a little mold sealer putty, or a rubber plug. Sometimes the molds come with a screw that can be screwed into the wick hole to help wax from coming out. I've found these to be unreliable, and prefer topping it with the mold sealer putty.

2. Secure the wick string at the top of the mold around a stick or metal wick bar.

3. Once wax reaches 180°F, add your fragrance (it's helpful to use a scale for a precise pour). You'll want to add about 1 ounce of fragrance for every pound of wax.

4. To create the dye effect, snap or cut the dye chip in half or thirds and add to the pot with the wax. You'll need to stir vigorously in order to fully disperse all the dye. Once it's mixed, you can test the color of the wax by dripping a little bit on a sheet of paper.

5. Once your wax has reached 140°F, pour your first layer. I always do a tepid pour first, just on the off chance my mold has decided to leak that day. If you want perfectly proportionate layers, use a scale and a smaller Pyrex pitcher to weigh just the amount of wax you'll need, repeating the same amount of wax for each layer. Otherwise, you need to be OK with some wabi-sabi levels until they create clear molds.

One note when you're working with molded candles—the bottom of the mold where you see a hole is actually going to be the **top** of the candle.

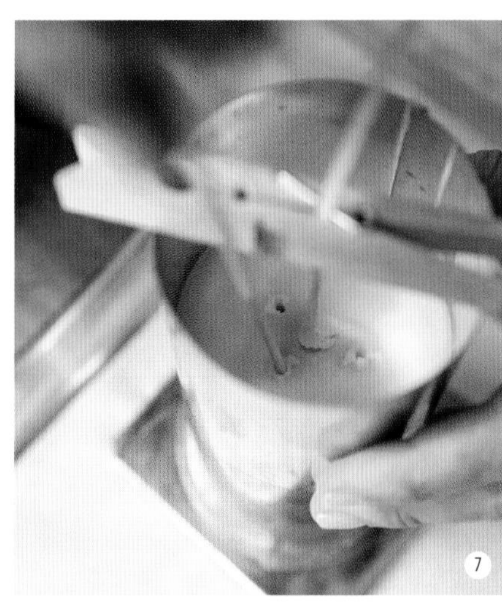

6. It will take about 45 minutes for your candle to set. Once it has, you'll need to create relief holes. Relief holes ensure there are no air pockets in the candle, especially along the wick, and on your last pour, they create a smooth bottom by filling in the well that naturally occurs as your wax contracts. Once the top layer has cooled, give it a poke with a skewer. Create three to four relief holes. Heat up your reserved wax, and once it's fully melted, add another small chunk of the dye chip. Pour the wax into the crevices you created with the stick. Allow this layer to cool completely before pouring the next layer.

7. The next layer is the same as the last. Add your last bit of dye chip and stir vigorously. Create relief holes in the previous layer of wax, and fill the cratered wax with your last bit in the pitcher, reserving only the smallest amount of wax for the final pour.

8. Once you've reached the desired height of your candle, finish with one last layer for a smooth bottom. If you see a crater form around the wick, make sure to poke a hole for an air bubble before the wax solidifies, and fill this in.

9. If you really want an even bottom, take an old baking tray and heat it up over the double-boiler system. Run the candle bottom along the tray, essentially melting the bottom and making it smooth.

10. Trim your bottom wick flush with the candle. Trim your top wick to ¼ inch.

DIFFUSERS

Diffusers are the lowest-maintenance when it comes to home fragrance—the classic set-it and-forget-it type. Diffusers release fragrance into the air using some sort of diffusing device, like bamboo sticks or electronic steam diffusers. It's a pretty simple concept: just add fragrance oil and a base together. The base is a stable material that keeps the fragrance from evaporating in a day or two, while also keeping the fragrance sticky.

We first started making diffusers because of a major account, CB2. CB2 was in love with our fragrances and wanted to know if we could make diffusers. It wasn't hard to make the jump, and Kristen designed the line using amber bottles to keep our apothecary aesthetic going. Diffusers have since taken off for us and are one of our personal favorites for fragrancing our home. We included a project for our favorite type of diffuser, a classic reed diffuser, in the following pages.

If you're into electronic steam diffusers, you can use one of the essential oil blends on page 108 and mix it with water. We use a steam diffuser in our daughter's room every night, and she does the duty of putting in the essential oil mix and filling the water. Her favorite scent right now is a blend of bergamot and lavender.

One Small Step Can Change Your Life
THE KAIZEN WAY
MAURER
workman
the perks of being a wallflower
STEPHEN
Signet Classics
GIOVANNI BOCCACCIO
THE DECAMERON
8.95 U.S.
(10.99 CAN)
AN AUTHORITATIVE ENGLISH LANGUAGE TRANSLATION OF THE MEDIEVAL MASTERPIECE
GIOVANNI BOCCACCIO

CALIF

REED DIFFUSER

WE USE REED DIFFUSERS IN JUST ABOUT EVERY ROOM of the house. Set in the corners and nooks of bathrooms, bedrooms, living rooms, and dining rooms, they fragrance the air subtly. You can customize the amount of fragrance being released by reducing the number of sticks—I usually keep around three for a smaller space, like a bathroom. If you need to freshen up the scent, flip the reeds.

The fun thing about reed diffusers is you get to use any container you like—a true combination of scent and interior style. You just need to make sure it's a nonporous material, since you don't want the container to absorb your diffuser liquid.

—KRISTEN

YIELD

MAKES 1 REED DIFFUSER

SUPPLIES

SCALE

GLASS OR METAL MEASURING CUP

METAL PITCHER

STIR STICK OR SPOON

METAL FUNNEL

GLOVES

SAFETY GLASSES

APRON

MATERIALS

CONTAINER

2¼ OZ FRAGRANCE

¾ OZ DPG, OR DIPROPYLENE GLYCOL, A TYPE OF ALCOHOL THAT DILUTES THE FRAGRANCE OIL AND HELPS IT WICK UP THE REEDS

RATTAN REEDS

1. Choose the container you want to use for your reed diffuser. For this project we are making a 3-ounce diffuser, so you want your container to be a bit larger to accommodate the displacement of liquid when the reeds are put into the container.

2. Use the scale to measure out the fragrance and DPG. Add your fragrance and DPG together into a glass or metal measuring cup.

3. Mix together for over 2 minutes to make sure the mixture is fully diffused. It is very important to do this in a metal pitcher or a glass container because the diffuser base/ fragrance combination can strip petrochemicals in plastics and paints, so it could ruin the finish of a plastic bowl.

4. Pour the mixture into the containers. We use a metal funnel to make sure that there is no spilling.

5. When ready to use, add the reed sticks to the container. It will take a couple of hours for the liquid to travel up the reeds and diffuse into the air. Rattan reeds are a natural product and they are prone to clogging with dirt after a while. A way to keep the diffuser scent strong is to flip the reeds to assist with the diffusion. As a business, we advertise that reed diffusers last three months, and, truly, that is their life with optimal performance. In more humid environments, like the bathroom, they can last a year or more.

Essential oil misters, like the ones pictured at right, are a quick way to experiment with scent daily. All they need is water and essential oils—we use them at night before bed. Use the essential oil blends on page 108, or experiment with your own mixes.

ROOM SPRAYS

Most people are familiar with room sprays because of commercially available products like Febreze, which is a favorite of parents and potheads alike. However, room sprays can be infinitely more sophisticated, scent wise, than being just a "laundry" style scent. The secret to an effective room spray is that the spray needs something to land on, preferably fabric.

Fabrics are great at soaking in scents—both good and bad. Dog odor, sweat, dust—these things build up and can make your sofa or pillows smell a little funky. Luckily, fabrics also absorb good scents well, too, so spraying a room spray on your pillows will quickly take care of that dog smell (I know it's not just us).

This is one of our favorite and most used projects from the book. Packaged in sleek bottles, we keep them on my nightstand and next to the couch to freshen up pillows or create a soothing sleep environment. If you're looking for a commercially available one, check out Fat and the Moon's *Calm Kid Mist*.

ROOM SPRAY

LIKE DIFFUSERS, ROOM SPRAYS ARE THE SIMPLE ADDITION of fragrance to a base material. Because room sprays are often applied to fabrics, like blankets and pillows, you don't want your base to be something that will stain a fabric, like an oil. And while water is a good carrier for fragrance, water and oils don't like to mix. To fix this, add an emulsifying agent to blend the two materials together—we used alcohol here.

In order to make room sprays for the P.F. line, we collaborated with our fragrance supplier to create water-compatible, body-safe versions of our scents, but you can create one more simply by using essential oils. The natural scents can be so light and mellow on a pillow. These essential oil room sprays can also be spritzed on the potpourri or sachet projects.

YIELD

MAKES ONE 2 OZ BOTTLE ROOM SPRAY

SUPPLIES

DROPPER OR PLUNGER

METAL FUNNEL

MATERIALS

60 DROPS (OR ~2 ML) ESSENTIAL OIL MIXTURE (SEE EXAMPLE IN STEP ONE)

2 OZ AMBER BOSTON JAR WITH SPRAYER HEAD

60 DROPS (OR ~2 ML) ALCOHOL (VODKA WORKS)

DISTILLED WATER

1. First, we will mix together essential oils to create a fragrance. You can use one of our suggested blends (see page 108), or try a simple ratio where about half of the mixture is the mid-notes, with the rest split between top and base notes. For example, if you mix lavender, rose, and cedar essential oils, you would do a mix of 30% lavender, 50% rose, 20% cedar. This can be mixed directly in your 2-ounce bottle.

2. In order to get the oil to mix with the water, we need to create an emulsification (it's a little bit like making a vinaigrette). In a jar or bottle, add 60 drops (2 ml) of your EO blend. Next add the exact same amount of alcohol (2 ml).

3. Mix together the essential oil and alcohol until emulsified. It can take a decent amount of mixing.

4. Once emulsified, transfer the mixture to the 2-ounce bottle using the funnel.

5. Fill the rest of the bottle with distilled water.

6. If the emulsification breaks, all you have to do is stir it up, or shake the bottle to get it to mix all back together. It's best practice to shake the bottle before each use.

INCENSE

I love incense. Always have. My love of incense might have started during my early twenties, when I would burn Nag Champa by the box to cover up whatever weird smells were bothering my roommates at the time. (That's also probably why I'm always lobbying for our vault scent Nag Champa to make a comeback.) Years later, it's still my number-one home fragrance tool.

In 2015, after years of working in production, we finally grew the company to a size where I didn't have to make every single unit that we were selling. It was an interesting time because I had to figure out what was next for me. I spent the next three months trying to develop new products. I bought a mortar and pestle, a sifter, and a ton of raw incense materials, and spent my days grinding plants and dried herbs to experiment with scent. It probably wasn't the best use of my time looking back, but it solidified my love of incense and let me experiment with fragrance in ways I never had.

To me, incense is the most rudimentary form of space fragrancing. Essentially you take some sort of plant material and you burn it. Is it rose petals or resin on a charcoal disc? Is it ground-up woods and plants shaped into a cone or block? Are you just straight-up burning the material like palo santo or sage? I always like to think that incense was invented when someone was trying to make a campfire, couldn't find any good wood to burn, so they threw some aromatic materials on it, and thought, "Hey, that smells nice."

Incense is a fragrance event. You have to be mindful while you light it and gently blow it out, since the smoke creates an atmosphere. The ritualistic aspect is why it's been used in so many religious ceremonies across cultures, from Catholic to Ancient Mayan practices.

It's easy to understand how incense fragrance works, as smoke creates a binding effect. Ask anyone who has been to a Las Vegas casino—you instantly smell like a cigarette for the rest of your trip. Smoke scent clings quickly and it is strong. That's why it's a great tool to quickly scent any room. In our house, we always light incense before we have guests over.

There are two types of incense you can make: loose (or non-combustible) incense, or combustible incense (like cones and sticks). These types are identical except for one crucial component—the smoldering agent. Cone and stick incense can be self-lighting, or combustible, with the addition of a smoldering agent, like charcoal or Makko powder. Non-combustible must be burned on a charcoal disk, or in the case of our campfire incense, in a roaring fire.

—TOM

LOOSE INCENSE

LOOSE, OR NON-COMBUSTIBLE, INCENSE IS the perfect place to start on a journey into incense making. It isn't as mainstream, but it's incredibly straightforward: dried materials burned on something hot, like a charcoal disk.

The most common way to burn loose incense is on a simple charcoal disk. You can also explore the traditional Japanese method of lighting incense by burying charcoal in white ash and placing a mica disk on top, called *kōdō*. This delicate method allows the incense to heat up, but not actually burn.

Loose incense is a great stepping stone to cone incense, because you can experiment with how plants, woods, and resins smell (or how they smell in combination) before grinding them into a powder.

My favorite material to burn on a charcoal disk is big chunks of resins like copal, amber, and frankincense. They're super smoky, so make sure you have a window or door open if you're doing it inside.

Safety tip with charcoal: These disks get insanely hot. Make sure you're burning it on something ceramic and heat resistant—not your favorite table. And whatever you do, don't move it while you're burning it.

—KRISTEN

YIELD

1 DISK OF INCENSE

SUPPLIES

ELECTRIC SPICE OR COFFEE GRINDER, IF DESIRED (A MORTAR AND PESTLE OR A MANUAL SPICE GRINDER WORKS TOO, IT'S JUST MUCH MORE WORK AND TAKES MUCH MORE TIME.)

MIXING BOWL

JAR FOR STORING

CHARCOAL DISK FOR BURNING

MATERIALS

APPROXIMATELY ½ TO 1 CUP RAW PLANT MATERIALS, RESINS, AND GUMS, OF YOUR CHOOSING, LIKE LAVENDER, ROSEMARY, OR FRANKINCENSE

1. You may choose to grind up all of your ingredients into a loose powder, or keep them in chunks. If powdering, grind your ingredients in a spice grinder or with a mortar and pestle until a fine, even consistency is reached.

2. Take the ingredients and mix together in a bowl.

3. Place materials in a container for safe keeping until ready to smolder. The more airtight the container the better, that way the materials will stay fresh and potent for the incense application later.

4. When ready to burn, place a pinch of the dry incense blend on a burning charcoal disk.

CAMPFIRE INCENSE

BACK TO WHAT I BELIEVE IS THE ORIGIN OF INCENSE—sitting around a campfire, throwing plant material in, and experiencing the scent they give off. It's so simple and oddly satisfying, and when you're camping, it's a nice way to get your clothes to smell like something other than smoke.

There was a time when Kristen and I went camping and made a pretty vigorous fire. We were with a big group, so a few people thought it would be fun to throw some random camp debris, like broken branches, into the fire. I'm not sure which one of the bushes was thrown in, but it smelled so good that we all went looking for more to put into the fire.

To me, you don't have to stop at a campfire. If you have a fireplace in your home that is used to burn logs, you can burn fragrance materials as well. One of our favorite recent memories was spending part of the winter in Santa Fe. The house we booked provided an unlimited supply of piñon logs for their kiva fireplace. Every night we built a nice roaring fire to keep us warm. The smell of those logs was so magical: sweet, spicy, balsamic, and earthy.

The wood itself can be a fragrance when you're setting a campfire, but you can also throw in these little bundles. They burn up quickly, but they help set the mood. If you don't have time, you can throw in pre-made sage bundles. You can wrap them up or put them in a small paper sack and throw the whole thing in.

—TOM

YIELD

MAKES 1 BAG OR BUNDLE

SUPPLIES

PAPER BAGS

TWINE

MATERIALS

DRY PLANT/WOOD MATERIALS (SUGGESTIONS PAGE 163, OR SEE BELOW)

JUNIPER

CEDAR

SAGE

LAVENDER

PALO SANTO

PINE

ROSEMARY

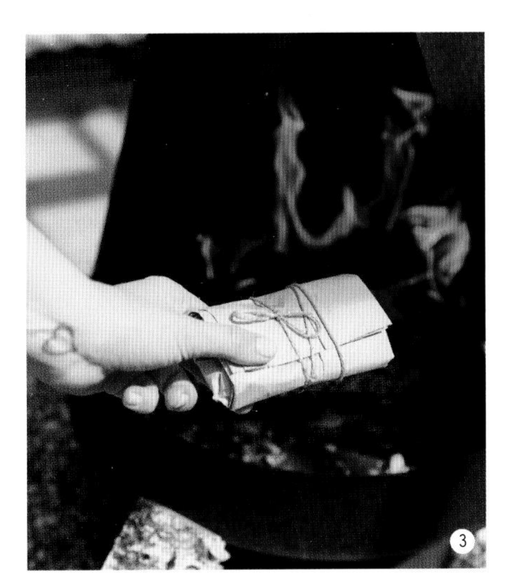

1. Gather materials needed for this campfire incense: Woods, sticks, and branches of bushes work especially well, but if you're putting it in a paper bag, you can also use materials like chunks of resin.

2. If making bundles, combine 8 to 10 sprigs of plant material together. Wrap the sprigs with twine. If using paper pouches, simply add your plant material to the bag, breaking down the pieces as needed to fit inside. Fold this into a smaller size (fire safety), and secure it with a piece of twine.

3. Once you have your fire going, you can carefully add the bag or bundle to the flames (using tongs) and enjoy the natural scent.

INCENSE CONE

INCENSE CONES ARE ONE OF MY FAVORITE THINGS TO MAKE. Once you get the hang of it, rolling and shaping the little cones is a nice meditation of its own. Incense cones have to fully dry before you burn them, so allow a week or two for drying time.

The main difference between loose incense and cone (or stick) incense is that you need to add a smoldering agent (in the case of the loose incense, the charcoal disk *is* the smoldering agent). You can use charcoal powder, but frankly it's really messy; it's such a fine particulate that one breath in the wrong direction leaves you covered in a black powder. Instead, we like to use Makko powder.

You're going to need anywhere from 25 to 70 percent smoldering agent to plant material, depending on what your blend is. Resinous materials, like copal and frankincense, require more Makko powder in order to burn correctly.

When I first started making incense, I felt overwhelmed by a lack of specific formulas, and then I realized that it's totally up to me to experiment. Working closely with natural materials really teaches you firsthand knowledge about these scents, as well. I've put a few of my favorite blends on page 163 as a jumping-off point. In the directions, I've referred to the ingredient measurements as "parts"—I used a teaspoon to measure my "parts," but you can certainly use a bigger measurement if you'd like to make more cones. Most of the blends on the following pages will make about six incense cones, if using a teaspoon as a "part." Add 1–2 more parts of Makko powder than the blends call for—the cones need the extra combustion.

All of your plant materials need to be finely powdered in order to roll into a cone. You can purchase most of these pre-ground up, which is a great time saver, or you can use an electric grinder (like a coffee grinder) or a mortar and pestle, if you want extra credit.

For both the incense cones and the sticks, you'll be working powdered plant materials and Makko powder, the smoldering agent, into a doughlike consistency.

—KRISTEN

YIELD

MAKES ROUGHLY 6 CONES

SUPPLIES

ELECTRIC SPICE OR COFFEE GRINDER, FOOD PROCESSOR, OR MORTAR AND PESTLE IF DESIRED, FOR GRINDING YOUR MATERIALS

1 TO 2 MIXING BOWLS, FOR POWDERS AND WATER

TRAY TO CONTAIN THE MESS

OPTIONAL: PIPETTE OR DROPPER

MATERIALS

POWDERED OR RAW PLANT MATERIALS AND RESINS, LIKE ROSEMARY, CEDAR, FRANKINCENSE, OR LAVENDER (SEE INGREDIENT INSPIRATION AND REQUIRED AMOUNTS ON PAGES 163)

MAKKO POWDER

WATER

1. The first step is to grind up all your plant materials. You can use a food processor, coffee grinder, or a mortar and pestle. I wouldn't recommend using anything you cook with, however, because residue from the natural materials may remain. The finer your grind, the easier it will be to roll and burn your incense later. Keep in mind that grasses and pine boughs don't grind very well, so they require extra time.

2. Mix together the powdered plant materials into a bowl.

3. Add Makko powder or a smoldering agent to your bowl. A minimum of 25 percent of your recipe total should be Makko powder, and up to 70 percent for highly resinous materials, like frankincense, amber, or copal. If using the formulas on 163, add 1–2 more parts of Makko powder for cones.

4. Add water to your bowl. You're looking to create a doughlike consistency here, like clay or cookie dough—just wet enough to hold shape while you mold it, but not crumbly. A pipette dropper is not necessary, but is immensely helpful to dole out small doses of water. I recommend working the dough with your hands, rather than a spoon. You'll need to add water until the dough starts coming together, like pasta dough.

    a. One technique to try is alternating Makko and water. You'll add one part Makko, one part water, one part Makko, one part water, etc. This inches you closer to that dough consistency and is helpful when you're first getting a hand of it.

5. Take off a small pinch, about the size of a small raspberry or grape, and roll this between your fingers. Shape the dough into a pyramid. To smooth out the edges, gently roll the sides out on a flat surface. Then, stand the cone upright, smushing it down ever so slightly to form a solid base on the bottom.

   Tip: If your incense is crumbling as you're molding it into cones, add more water. If it's sticking to your hands, add more Makko. You can also use a baking technique with your hands and wet your hands with water, or dip your fingertips in Makko powder as you work the dough, depending on if you need it more wet or dry. This prevents you from adding too much of either component.

6. Let your incense dry for one to two weeks before you burn it. This may be shorter depending on your environment (ours dry much faster here). A wet incense may light, but won't smolder correctly for a full burn.

7. Burn it! Find a heat-resistant surface to place your cone on. Light the top of the cone and, when it's clear it is lit, gently extinguish the flame and let the cone smolder. Each cone will burn for about 25 minutes. If your incense won't burn, it could be that it needs more time to dry, or that it needs more smoldering agent. If it is the latter, you can always burn incense on a charcoal disk.

ROLLED INCENSE STICKS

ROLLED STICKS USE THE BASIC PREMISE OF CONE INCENSE, but instead of molding the dough into a cone shape, you roll it onto a bamboo stick (or skewer, if you can't find those). Incense sticks have a pretty steep learning curve, so be patient with yourself. The key for incense sticks is to try and try again. Cutting your bamboo skewer in half also makes it more manageable to work with. These incense sticks are going to turn out a little chunky, but I've found they still dry faster than cones—so although they're harder to make, they're easier to burn.

—KRISTEN

YIELD

MAKES APPROX. 3 STICKS

SUPPLIES

ELECTRIC SPICE OR COFFEE GRINDER, FOOD PROCESSOR, OR MORTAR AND PESTLE, IF DESIRED, FOR GRINDING YOUR MATERIALS

1 TO 2 MIXING BOWLS, FOR POWDERS AND WATER

OPTIONAL: TRAY TO CONTAIN THE MESS

OPTIONAL: PIPETTE OR DROPPER

BAMBOO STICKS OR SKINNY SKEWERS

MATERIALS

APPROX. 5 TSP OF POWDERED OR RAW PLANT MATERIALS, GUMS, AND RESINS, LIKE PINE, MYRRH, YARROW, OR SAGE (SEE INGREDIENT INSPIRATION AND REQUIRED AMOUNTS ON PAGE 163)

MAKKO POWDER (YOU'LL WANT AT LEAST 25% OF YOUR TOTAL MATERIALS TO BE MAKKO—MORE IF YOU USE A LOT OF GUMS AND RESINS)

WATER

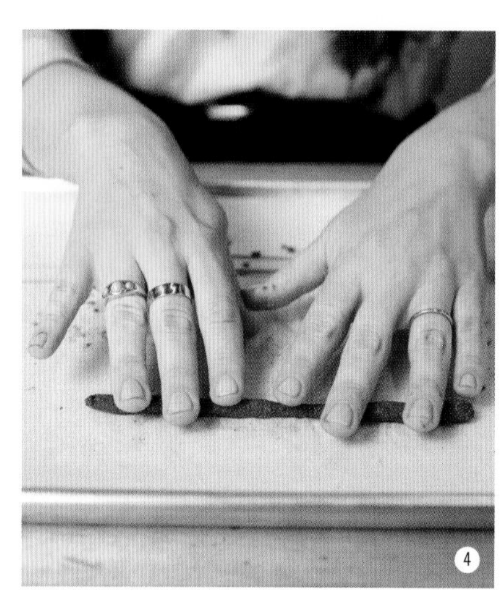

1. The first step is to grind up all your plant materials. You can use a food processor, coffee grinder, blender, or mortar and pestle (don't use one that you'll use for food later).

2. Mix the plant material and resins in a bowl, then add your Makko powder.

3. Add some water. Be stingy, as this dough works better when it's slightly sticky. You can always add more water, but it's harder to go back once you've added too much. A pipette is very helpful when adding water.

4. You'll first need to make a long, thin shape with your dough. To do this, you can create a "snake" by rolling the dough between your hands. You'll want to create a shape that is 1 to 2 inches shorter than your overall stick length, about ¼ inch wide and ⅛ inch tall.

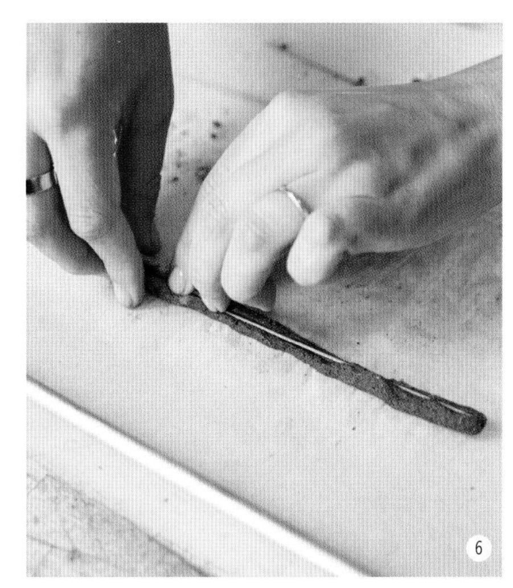

5. Push the stick down into the mound, allowing the dough to squeeze up around the sides.

6. Pinch the dough in place around the skewer, insuring full coverage.

7. Now you need to smooth the dough onto the stick. Roll the dough-covered stick like you're using a rolling pin—but be gentle. This is the trickiest part of this process, as sometimes the dough will fall off or become too heavy, so you may have to start over. Starting over isn't a big problem, though—it will work your materials into a finer particulate. If your dough starts to break, try wetting your fingers and smoothing out the cracks.

8. Continue rolling, smoothing, and pinching the dough until it is an even consistency on the stick, with about 1 inch left at the bottom. If you're having issues with the rolling pin method, you can try twisting and pinching the dough down the stick, then smooth out

9. Let your sticks dry for at least a week before burning.

Incense Blends

The incense in this book is made using raw plant and natural materials, like lavender sprigs, hunks of amber and frankincense resin, and palo santo. The blends that follow are general scent profiles that can be applied to any of our projects, whether it's campfire incense or rolled sticks. The main thing that will change is whether the material is powdered or left in whole chunks. For the campfire incense, the materials can be left whole to burn, or you can use powders. For powdered or stick incense, you'll want to achieve a finely powdered consistency, like flour. The Makko powder and water are only needed in the cone and stick incense formulas.

MATCHA BLEND

I CALLED THIS BLEND my matcha blend because of the bright green color of the sweetgrass. Sweetgrass is one of my favorite burnables because it imparts a warm, vanilla sweetness when it's burned.

---

2 PARTS SWEETGRASS
1 PART LAVENDER
1 PART FRANKINCENSE
2 PARTS MAKKO POWDER
2½ PARTS WATER

SWEETHEART BLEND

CRUSHED ROSES MAKE UP the heart of this blend. This is a great beginner choice because there's not a lot of resinous materials, making it easier to work with and quick to blend.

---

2 PARTS ROSE POWDER
1 PART SAGE
1 PART PALO SANTO
2 PARTS MAKKO POWDER
2 PARTS WATER

OJAI BLEND

THIS RECIPE WAS INSPIRED by wildflowers and rolling chaparral hillsides near Los Angeles. Lavender is the real star here.

---

2 PARTS LAVENDER
2 PARTS YARROW
1 PART FRANKINCENSE
2 PARTS MAKKO POWDER
~2 PARTS WATER

SPICY WOODS

THIS SCENT FEELS LIKE a real P.F. scent: woodsy with a touch of spice. You can substitute juniper or another type of wood here, or add sweetgrass, which would give it a New Mexico vibe.

---

2 PARTS PINE
1 PART CEDAR
1 PART CINNAMON
2 PARTS MAKKO POWDER
4 PARTS WATER

Note: when using these blends for cones, add 1–2 extra parts of Makko powder.

POTPOURRI

When I think about potpourri, what immediately comes to mind are those noxious, dried flower petals in wicker baskets that were ubiquitous in the nineties. They came pre-packed from decor stores and didn't really trend past 1999. You're probably surprised to see an entire section on potpourri here.

Recently, I've been thinking of what potpourri actually is and how interesting dried scented material can be as a home fragrance. After all, it's just dried herbs and flowers placed together to scent your house. When I think about it, it is not too dissimilar from incense, where we are using natural materials for their fragrance. Nowadays, people still keep dried herbs and florals around—they're having a moment—but typically in the form of arrangements or dried sage bundles.

On a recent trip to Boston, our friend and P.F. stockist, April, of Boston General Store, tasked me with telling her exactly what her shop smelled like. Visitors to her shop frequently ask her, "What smells so good in here?" and while I'd love to take credit, since she stocks almost our entire line of candles and diffusers, I did some detective work and think that a big contributor is actually the dried materials, like straw bags and baskets, lavender, herbs, and bath salts that she stocks.

Basically, I'm saying to upend your thinking on what potpourri is. Organic and natural materials, foraged from a walk outside or in the park, are frequently on display in our house in the form of rocks, driftwood, seed pods, and flowers. These dried materials can work double duty with the application of fragrance and fixative.

The secret with potpourri is that the dried materials aren't going to contribute as much fragrance to the end result as you think, unless you're like April and have an entire store full of dried goods. The fragrance and fixative is key for getting the scent to last.

We've included a basic fixative recipe, a list of materials, ideas, and steps to follow to create your own potpourri, but the inspiration is truly your own here. If you don't have access to nature where you are, dried materials can also be ordered online.

You can also spray the essential oil room spray on your potpourri.

—KRISTEN

BOWL POTPURRI

POTPURRI IS SIMPLE TO MAKE—it's up to you to pick the materials that suit your aesthetic and aren't too fussy. I love the look of lotus pods, found at the Los Angeles flower market, and combined them with pomegranates from the tree growing in our yard and with dehydrated apple slices. This is a pot I'd display in the fall—it's not too on the nose with leaves or pumpkins, but still brings a little seasonality inside.

To dry materials like orange and apple slices in the oven: Set your oven to the lowest setting. Cut your fruit of choice into thin slices, then place on a baking tray in the oven. Bake for several hours, or until fruit has reached the dried consistency you like.

YIELD

1 BOWL

SUPPLIES

PAPER BAG (COULD ALSO USE A SEALED CONTAINER OF SOME SORT)

DECORATIVE BOWL OR DISH

MATERIALS

DRIED MATERIALS LIKE FLOWERS, HERBS, SEED PODS, PINE CONES, OR SPICES

FIXATIVE (POWDERED ORRIS ROOT)

ESSENTIAL OILS

1. Dry out any florals or leaves and trim off the stems.

2. Combine 2 to 4 cups (depending on the size of your end vessel) of dried materials into the paper bag or sealable container.

3. Add 1 to 2 teaspoons of powdered orris root with about 15 drops of your favorite essential oil to the container. If you like it stronger, experiment with more drops of EO. Shake to combine.

4. Leave the bag or container sealed for 2 to 3 weeks before using to fully cement the fragrance onto the dried materials.

5. When you're ready to use, empty the contents into a nice wood bowl or ceramic dish.

DRAWER SACHETS

I GOT INTO SACHETS ON A TRIP TO WASHINGTON, D.C., where we were visiting Kristen's parents in Northern Virginia. It was a rainy day, but we decided to take the train into D.C. to visit museums (yay for free museums). We got soaked walking around the National Mall and warmed up wandering around the museums—and their gift shops. In one gift shop, we found a sachet printed with a bright, Southwestern pattern, stuffed full of lavender and sage. For years we hung that sachet on our bed frame, poking it every so often to release the scent. The bright pattern and mix of fragrant materials upended our expectation about what a sachet could be, which previously seemed like fussy or purely functional affairs.

Sachets kept in drawers also have the benefit of keeping away moths. Lavender, mint, and cedar can all keep pests away.

These are a cinch to make using pre-sewn muslin bags. You can even use some of your potpourri and bag it up—especially useful if your potpourri has gotten crumbly.

—TOM

YIELD

MAKES 1 SACHET

SUPPLIES

MUSLIN BAGS

5 CLOTHESPINS

SCISSORS

MATERIALS

1 CUP DRIED MATERIALS (SEE PAGE 175), BUT LAVENDER, SAGE, CEDAR, AND EVERGREEN ARE ESPECIALLY GOOD

ESSENTIAL OIL OF YOUR CHOOSING

ORRIS ROOT (OPTIONAL)

VENDER
1

1. Gather your dried materials, making sure they are all properly dried.

2. Combine together ½ to 1 cup of dried materials, depending on how much fits in your muslin bag, in a bowl or straight into your muslin bag.

3. Add about 15 drops of your favorite essential oil or essential oil blend. Add 1 teaspoon of orris root as a fixative, if desired.

4. Decant into a muslin bag, draw up the strings, and you're ready to use!

Potpourri Blends

A few blends to get your mind going.

Summer Blend

Chamomile, daisy, or other
white/yellow flowers
Eucalyptus, seeded if you can find it
Lotus seed pod

Holiday Blend

Pine cones
Dried orange slices
Cinnamon sticks

Fall Blend

Small dried gourds
Sliced apples
Seed pods

DRIED MATERIAL IDEAS

Flowers, dried

Chamomile

Roses (buds work especially well)

Carnations

Daisies

Wax flowers

Herbs and plants

White sage

Pine

Juniper

Sweetgrass

Eucalyptus

Spices

Cinnamon

Star anise

Cardamom

Seed pods, like lotus root or gum pods

Pine cones

Small dried gourds

SIMMER POTS

SIMMER POTS HAVE GAINED IN POPULARITY over the past couple years since they're a natural and organic way to bring fragrance into your home, with minimal waste. The basic premise is to simmer fruits, herbs, and spices in a small pot of water on the stove, and is definitely inspired by mulled cider or massive cooking days, when your whole home is enveloped in gourmand scents. For us Americans, no day has a bigger gourmand scent memory than Thanksgiving—all day cooking poultry or pork; creamy, savory, and buttery sides; spicy and sweet pies and cookies—it's a true scent event.

We've had our fair share of warm ninety-degree Thanksgivings in Los Angeles, but when it is cold and dreary, those scents really pick up our spirits, and there's something about the low burble of a pot that makes you feel cozy, like you're tucked inside on a rainy day with a cup of tea.

Simmer pots really bring me back to our first house together in Austin. That place was just a glorified two-bedroom shack, with zero insulations on the walls or floors (you could actually see between the cracks in the floorboards to the ground below), and whatever temperature it was outside, so it was on the inside. It wasn't the worst, because our rent was eight hundred dollars a month. Austin is known for being hot, but because of its location just south of the Midwest, these Canadian cold fronts would come down, bringing the temperature down to between fifteen and thirty-five degrees, one to two times a week in the winter. During our first winter there, we didn't understand how to get this archaic 1950s wall heater to work, so we used our stove to heat the house with two methods. First, we left the oven cracked with a beer can to heat the house (don't try this at home—safety first). Second, Kristen would make candles on the stovetop. This was when she used the double-boiler system, so it was essentially a simmer pot. The slow sound of boiling water, the humid air boiling out of the pot, and the smell of gourmand fragrance oils would fill the air. We were broke, but I always look back at that time with such fondness, because at least we did the best we could.

Simmer pots work best with a juicy fruit or citrus, spices like cinnamon or clove, or aromatic greens like eucalyptus or mint. We've given some recipes for our favorite simmer pots below, and also a list of ingredients to try. There's not really a way to go wrong here—if you put something stinky in there (like we did, with fennel—not our bag), you can just fish it out and keep on simmering. Simmer pots are also a great way to use produce that's gone bad in your fridge.

—TOM

YIELD

MAKES 1 SIMMER POT

SUPPLIES

POT

STOVE

MATERIALS

PLANT MATERIALS (SEE PAGE 175 FOR SUGGESTED COMBINATIONS)

WATER

1. Slice up your fruits and pull apart leaves of herbs (this allows the water to simmer around your materials more).

2. Add the ingredients to the pot.

3. Fill the pot with water, stopping an inch or so below the lip of the pot. This allows headspace so that the pot doesn't overflow when it simmers.

4. Turn the burner on low and simmer away. You may choose to boil first and bring to a simmer, but make sure you are watching the pot just in case it boils over.

5. The simmering pot will fragrance the house. Keep an eye on the pot to make sure the water doesn't boil down too much. Refill when the water has lowered.

   Tip: Alternatively, you can do this in a slow cooker. Add materials and water to the slow cooker, put on low, and let the pot simmer the fragrance materials.

Simmer Pot Blends

Fall Pot

Apples (sliced)
Oranges (quartered)
Cinnamon sticks
Cardamom

Spring Pot

Eucalyptus
Orange
Lavender

Summer Pot

Mint
Basil
Strawberry

Winter Pot

Ginger
Rosemary pine
Lemon boughs

SIMMER POT MATERIAL IDEAS

Apples
Oranges
Lemons
Grapefruit
Eucalyptus
Geranium
Lavender
Basil
Peppermint
Rosemary
Chamomile
Mint
Pine
Cinnamon
Ginger
Cardamom
Star anise
Clove
Black peppercorns

Resources

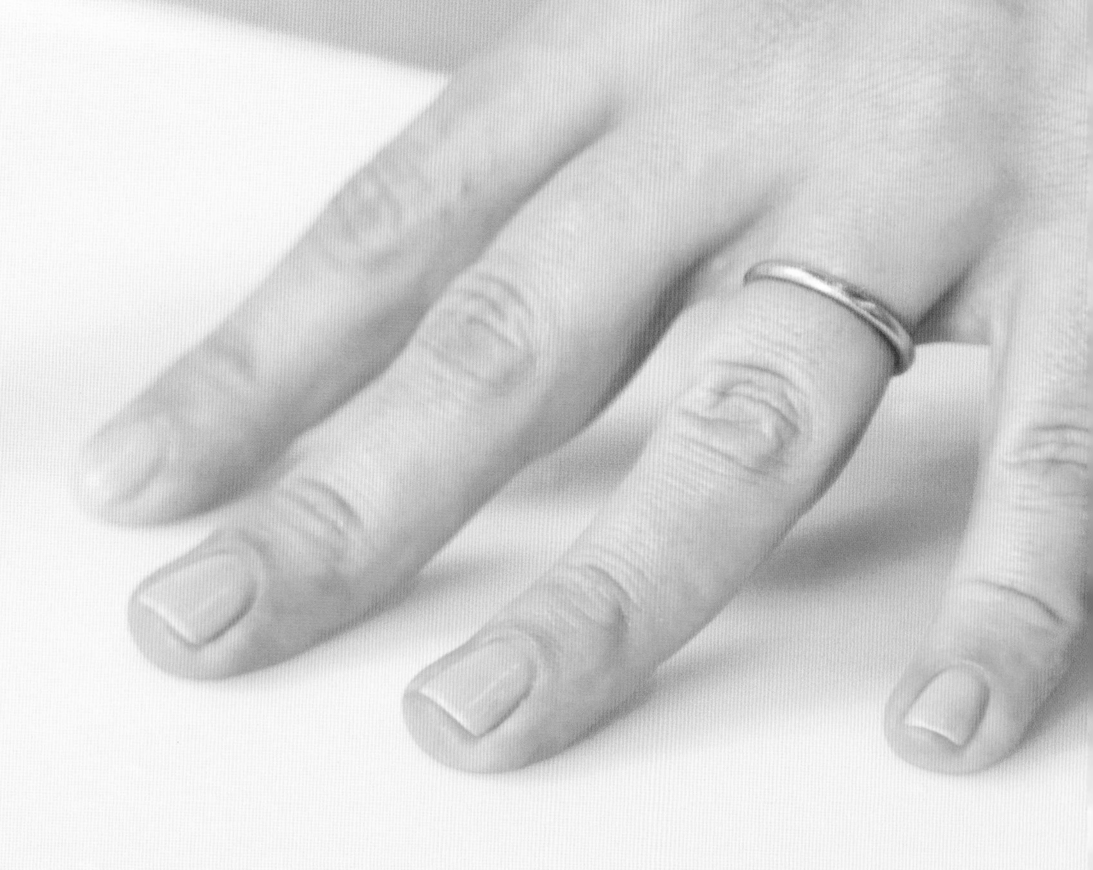

California Candle Supply

calcandlesupply.com

FRAGRANCE, CONTAINERS, CANDLE MAKING SUPPLIES (WICKS, DYES, WAX, CONTAINERS), EQUIPMENT, CANDLE MAKING KITS

Candlewic

candlewic.com

CANDLE MAKING SUPPLIES (WICKS, MOLDS, DYES, WAX), SOAP MAKING SUPPLIES

CandleScience

candlescience.com

FRAGRANCE OILS, CANDLE MAKING SUPPLIES (DYE, WICKS, MOLDS, WAX),
SOAP MAKING SUPPLIES, DIFFUSER MAKING SUPPLIES (DIFFUSER BASE AND REEDS), PACKAGING, EQUIPMENT (PITCHERS, THERMOMETERS, WICK BARS, WICK SETTERS, WAX MELTER, WICK STICKERS)

Craft Supply Store

Local craft supply store or chain (Michaels or Jo-Ann Stores)

CANDLE MAKING KITS, FRAGRANCE,
ART SUPPLIES, CERTAIN KITCHEN SUPPLIES

Flower Store

Local flower market or shop

FLORALS, DRIED MATERIALS

Grocery Store

Local grocery store or farmers market

HERBS, SPICES, FLORALS, FRUITS, MAKER SUPPLIES (POTS, THERMOMETERS, FUNNELS, WOODEN SPOONS, CLOTHES PINS, SPICE GRINDERS, MORTAR AND PESTLE), TIMER

Hardware Store

Local hardware store or national chain (Ace Hardware, Home Depot, Lowe's)

SCALES, PROTECTIVE GEAR
(GLOVES, EYE GOGGLES)

Lone Star Candle Supply

lonestarcandlesupply.com

FRAGRANCE OIL, CANDLE MAKING SUPPLIES (WICK, WAX, DYES, CONTAINERS)
DIY KITS, EQUIPMENT (WICK BARS, POTS, WICKS SETTER, SCALE, THERMOMETER)

Mountain Rose Herbs

mountainroseherbs.com

DRY HERBS, SPICES, RESINS, GUMS, FIXATIVES, ESSENTIAL OILS, INCENSE MATERIALS, CHARCOAL, CARRIER OILS, WAXES, CONTAINERS, KITCHEN SUPPLIES (MORTAR & PESTLE, FUNNELS, SPICE GRINDER)

Rite Hete

ritehete.com

MELTING TANKS

Rustic Escentuals

rusticescentuals.com

CANDLE MAKING SUPPLIES (WICK, WAX, DYE, CONTAINERS), CANDLE MAKING EQUIPMENT (PITCHER, SCALES, MIXERS, PIPETTES, THERMOMETERS), SOAP MAKING SUPPLIES, PACKAGING, FRAGRANCE OILS

Save on Scents

saveonscents.com

CANDLE SUPPLIES (WAX, WICK, DYES, CONTAINERS), FRAGRANCE OILS, ESSENTIAL OILS, SOAP MAKING SUPPLIES, INCENSE MAKING SUPPLIES (UNSCENTED CONES AND STICKS), DIFFUSER SUPPLIES (DIFFUSER BASE, REEDS), EQUIPMENT (PIPETTES)

Scents of Earth

scents-of-earth.com

INCENSE MAKING SUPPLIES (RESINS, GUMS, WOODS, HERBS, BAMBOO BLANKS, MAKKO POWDER), INCENSE BLENDS, INCENSE ACCESSORIES, FIXATIVES, CHARCOAL

Wax Melters

waxmelters.com

MELTING TANKS (CANDLES AND SOAP), FILLING SYSTEMS (CANDLES, ROOM SPRAYS, DIFFUSERS, HAND SANITIZER)

Wholesale Supplies Plus

wholesalesuppliesplus.com

SOAP MAKING SUPPLIES, DYES, FRAGRANCE, ESSENTIAL OILS, CANDLE MAKING SUPPLIES, KITS

The Wooden Wick Co.

woodenwick.com

FRAGRANCE OILS, ESSENTIAL OILS, CANDLE MAKING SUPPLIES (WICKS, WAX, DYES, CONTAINERS), SOAP MAKING SUPPLIES, ROOM SPRAY SUPPLIES, DIFFUSER SUPPLIES (REEDS, DIFFUSER BASE), EQUIPMENT (PITCHER, THERMOMETER, SCALE, BLOTTER STRIPS), DIY KITS

Glossary

Absolute
An extraction method for materials. The material is soaked in a substance, and the fragrance is washed out.

Accord
A blend of fragrance materials to produce a new fragrance. Some accords will be created as a way to try to recreate a natural fragrance.

Animalic
These are fragrances that are derived or give the impression they are derived from animals. Some examples of derived fragrances are civet, ambergris, and castoreum. Costas root is from a plant, but smells like it could have come from an animal.

Aromatic
A quality of fragrance that opens up the nasal passages. Materials that are considered aromatic: mints, lavender, eucalyptus.

Balsamic
A resinous scent quality that is woody, sweet, and a little spicy at the same time.

Base Notes
The most sturdy fragrance molecules. The base notes are the part of the scent that lasts the longest.

Camphoraceous
Smelling of camphor. Camphoraceous fragrances are aromatic like eucalyptus, but are also a little funky.

Cooling
In candle making, cooling is the process of letting the molten wax turn into a solid form.

Distillate
What the byproduct of distillation is called.

Dry Down
The fragrance characteristic that is expressed after a fragrance material has had an opportunity to sit and dry.

Earthy
Smelling of soil, dirt, and earth.

Fixative
Materials that reduce the volatility of other fragrances. They tend to have base, not tenacity.

Green
Fragrances that smell of plant material such as grass and leaves.

Herbaceous
Expressing characteristics of herbs. Green, earthy, cool, leafy. To us, it is best experienced in combination with basil.

Indolic
Indol is a fragrance material that is the byproduct of decaying materials. To most people indol smells of halitosis, but it is also a major constituent of jasmine.

Linalool
Linalool as a green note on its way to be floral, that stops just short of the petals. Linalool is found in a ton of fragrance products—it's very popular—and in things like rose, neroli, and a large component of lavender. Linalool is sweet with just a touch of juicy citrus.

Middle Notes
Characteristics of fragrance that are in the middle of the tenacity spectrum. These are not the things that last a long time, but aren't volatile notes that quickly disappear.

Ozonic
Smelling of ozone. Ozone can best be described as the scent given off after a lightning strike.

Tenacity
Length of fragrance. Regarded as top, middle, and base. Top is the most fleeting, while base is strongest. Generally related to the stability and size of the molecules that the materials are made of.

Terpenic
A resinous fragrance material classified as terpenes. Turpentine is the strongest distilled representation of a terpenic fragrance. Terpenes exist in many materials such as citrus, pines, and cannabis.

Texture
How the material interacts with your nose. Is it powdery, scratchy, ambery, smooth, velvity, or cool?

Top Notes
These are the most volatile compounds of a molecule. Top notes are the part of the fragrance that is smelled first and does not linger.

Triplal
A fragrance material that is the definition of green. It smells just like a grass lawn in the middle of spring.

Woody
These are fragrances derived from wood and smell like it. They give the impression of bark and fresh-cut or sawed wood. Examples are pine and sandalwood.

CROSLEY
SEMI AUTO STOP
2-BAND RADIO CASSETTE RECORDER
SONY
harman/kardon 330B
POWER
PHONES

This book would not have been possible without the support of our staff, past and present. Thank you for working tirelessly by our sides to build this business, and for stepping up on our WFH Wednesdays so we could bring our book to life.

To the Development Team, Jade Meresz and Kim Dang, thank you for testing our formulas and providing critical feedback. To Kirsten Wiltjer, thank you for pushing along our core operations while we were out writing.

To Meredith Clark, our intrepid editor, who slid into our DMs one day and asked if we ever thought about writing a book: thank you so much for your thoughtful, kind, and encouraging guidance throughout this journey. Thank you to the rest of the Abrams team who worked on this project—there are so many unseen hands that pass over a book in order to bring it to market. To our talented designer Heesang Lee, the way you brought the P.F. world to the page was so fun to watch.

Grant Puckett, you're probably one of the hardest working photographers in the biz. Thank you for being an integral part of building our aesthetic over the years. Plus, doing the photoshoots in the tumultuous days before the quarantine is an experience we will never forget.

We would be remiss not to acknowledge the formidable Institute for Art and Olfaction, a non-profit perfumer's Institute in LA's Chinatown. Their incredible knowledge and training were instrumental in strengthening our language and translating scent onto the page.

To Poppy: our most incredible creation and the person who gives us purpose and motivation. We love to experience the world with you. Hopefully, every time you smell something, you'll think of us and know we'll be there for you forever.

From Kristen:

Mom and Dad, you empowered me and supported me, never losing patience while I tried my hand at various ventures, from ice skating to acting to running a business.

To my sister, Jenne, your angsty goth phase in high school taught me what it means to be an individual. If you had never bought that candle making kit for your Home Ec fair, I probably wouldn't be doing what I'm doing today.

Tom, through marriage, business, and parenting, you have shown me what a true partner looks like. This book would not have been possible without your project management, even though I'm sure you felt like you were herding a cat sometimes.

From Thomas:

To my Mom, thank you for giving me everything you had and thank you for giving me the freedom to figure out who I was. Thank you for showing me how to persevere in an unkind world.

My sister Tammy and my brother-in-law, Rick, thank you for being there when no one else could be, teaching me the value of hard work and appreciating the things that I have.

To my friends Bree and Casey, thank you for taking care of me that year when I was a mess. Without y'all, I would have never gone to The Long Branch Inn that night in August 2008 and met Kristen. None of this would be possible without that convergence of worlds.

Is this cheesy to acknowledge the other author? To my partner in crime, Kristen, thank you for giving me all the world's opportunities. I could fill the entire book with just my acknowledgment to you. Thanks for challenging and encouraging me all these years. I am excited for the next challenge we take on together.

Editor: Meredith A. Clark
Designer: Heesang Lee
Production Manager: Kathleen Gaffney

Library of Congress Control Number: 2020943999

ISBN: 978-1-4197-4627-7
eISBN: 978-1-68335-959-3

Printed and bound in China
10 9 8 7 6 5 4 3 2 1